F.H. Schweingruber   A. Börner   E.-D. Schulze

Atlas of Woody Plant Stems
Evolution, Structure, and Environmental Modifications

F.H. Schweingruber
A. Börner
E.-D. Schulze

Atlas of Woody Plant Stems
Evolution, Structure, and Environmental Modifications

With 710 illustrations, most of them in colour

Springer

Prof. Dr. Fritz Schweingruber
Institute for Forest, Snow and Landscape Research WSL
Zürcherstrasse 111
8903 Birmensdorf, Switzerland

Annett Börner
Prof. Dr. Ernst-Detlef Schulze
Max Planck Institute for Biogeochemistry
PO Box 100164
07701 Jena, Germany

Front cover illustrations (from left):
 Cross-section of a pollarded ash stem (*Fraxinus excelsior*). The light zones indicate small rings which are an expression of a pollarded crown. Valais, Switzerland.
 Cross-section of a petrified, fossil *Araucaria* stem. Petrified Forest, Arizona, USA.
 Cross-section of dead wood from an Red Bearberry (*Arctostaphylos uva-ursi*). Most lumina of the cells are filled with brown substances which prevent the wooden structure from microbiological degradation (magnification 100×).
 Cross-section of a birch stem (*Betula pendula*) with demarcation lines. They represent concentrations of mycelia of different fungus species.

Library of Congress Control Number: 2006927045

ISBN-10 3-540-32523-9 Springer Berlin Heidelberg New York
ISBN-13 978-3-540-32523-9 Springer Berlin Heidelberg New York

This work is subject to copyright. All rights are reserved, whether the whole or part of the material is concerned, specifically the rights of translation, reprinting, reuse of illustrations, recitation, broadcasting, reproduction on microfilm or in any other way, and storage in data banks. Duplication of this publication or parts thereof is permitted only under the provisions of the German Copyright Law of September 9, 1965, in its current version, and permissions for use must always be obtained from Springer. Violations are liable for prosecution under the German Copyright Law.

Springer is a part of Springer Science+Business Media
springer.com

© Springer-Verlag Berlin Heidelberg 2006
Printed in Germany

The use of general descriptive names, registered names, trademarks, etc. in this publication does not imply, even in the absence of a specific statement, that such names are exempt from the relevant protective laws and regulations and therefore free for general use.

Editor: Dr. Dieter Czeschlik, Heidelberg, Germany
Desk editor: Dr. Andrea Schlitzberger, Heidelberg, Germany
Cover design: Design & Production GmbH, Heidelberg, Germany
Camera-ready by Annett Börner, Jena, Germany
Production: LE-TeX Jelonek, Schmidt & Vöckler GbR, Leipzig, Germany

Printed on acid-free paper   31/3100/YL   5 4 3 2 1 0

Preface

The "Atlas of Woody Plant Stems" is a comprehensively illustrated book with short, informative texts. We chose this layout because plant anatomy and morphology can only be conveyed by detailed pictures. In addition, a vivid presentation should attract a broader public, not only the specialist. We hope that the combination of anatomy and morphology will create interest and curiosity. Amateurs will enjoy the wide range of pictures; interested readers will be caught by particular chapters; specialists will delve into aspects and photographs that may have never been presented before; teachers may use the pictures for illustrations in classes with students.

There is, to our knowledge, no other book that presents the concept "stems" in the combination of anatomy and morphology. So far, a work that explains the fundamental anatomical aspects of dendrochronology is lacking. Also, we are not aware of another book that combines the macroscopic and microscopic structures in wood and bark in branches, stems, and roots and of trees and herbs. We also believe that there is no book available that puts these structures into a chronological, morphological, ecological and taxonomic context.

Naturally, it was impossible to cover completely the enormous variability of plant life forms. We have tried, however, to illustrate the main principles and features. Many decades of collection and preparation provided the basis for this book.

Fritz Schweingruber would like to thank the Swiss Federal Research Institute WSL, that offered him hospitality after his retirement. The authors thank all the students and colleagues that have, for decades, collected and prepared samples, and who helped with the English translations. Thanks to John Kirby who made the final English editing.

Fritz H. Schweingruber, Annett Börner and
Ernst-Detlef Schulze
Birmensdorf and Jena, May 2006

Preface	V
Abbreviations and Technical Remarks	X
Introduction	1

1 The Evolution of Plant Stems in the Earth's History

The Landscape in the Paleozoic	4
Plant Body of Vascular Plants	6
The Evolution of a Stabilizaton System	8
The Contemporary Fossil *Psilotum Nudum?*	9
Diversification of Plants Containing Tracheids	
The Lycopods	10
The Horsetails	11
The Fossil and Contamporary Ferns	12
Contemporary Ferns	14
Trees Grow Taller and Bigger	16
Successful Seed Plants with Naked Seeds	
Ginkgos and Cycads	18
Gnetophytes (*Ephedra, Gnetum* and *Welwitschia*)	20
The Most Successful Seed Plants with Naked Seeds: Conifers	22
Successful Plants with Seeds Enclosed in a Carpel: Angiospermae	24
Systematic of Plant Life	26

2 The Structure of the Plant Body

Life Forms in Different Vegetation Zones	28
Principal Growth Forms of Stems	30
Principal Construction of Roots and Shoots	32
Principal Construction of the Xylem and Phloem	
Cell Types, Cell Walls and Cell Contents	34

3 Secondary Growth: Advantages and Risks

Primary and Secondary Growth	40
Principle Structure of Plants with Secondary Growth	42
Physiological Ageing in Plants with Secondary Growth	43
The Risks of Water Transport:	
Stabilized and Permeable Cell Walls	44
The Risks of Stem Thickening:	
Dilatation and Phellem Formation	46
The Risks of Over-Production:	
Programmed Cell Death	50
The Risks of Instability:	
Eccentricity	52
Reaction Wood	54
Formation of Lignin and Thick Cell Walls	56
Internal Optimization	58
The Risk of Decomposition:	
Natural Boundaries and Protection Systems	60
Defence Barriers Around Wounds	62
The Risk of Shedding Plant Parts:	
Abscission	64

4 Modification of the Stem Structure

The Primary Stage of Growth:	
The Construction of Vascular Bundles	70
The Arrangement of Vascular Bundles in Mosses, Lycopods and Ferns	72

    The Arrangement of Vascular Bundles in Conifer and Dicotyledonous Plant Shoots 74
The Secondary Stage of Growth:
  Conifer Xylem 76
  The Xylem of Dicotyledonous Angiosperms 78
The Primary and Secondary Stages of Growth of Monocotyledons:
  Macroscopic View 82
  Microscopic View 84
The Secondary Stage of Growth:
  Conifer Phloem 86
  The Phloem of Dicotyledonous Angiosperms 88
  Cambial Growth Variants and Successive Cambia 90
The Third Stage of Growth: The Periderm 92

5 Modification of the Xylem Within a Plant

Modification of the Xylem Within a Plant
  Conifer: Root, Twig and Stem 96
  Deciduous Tree: Root, Twig and Stem 98
  From Root to Stem Structure 99
Modification by Aging:
  Changing Growth Forms 100
  Changing Growth and Leaf Forms 101
  Changing Wood Anatomical Structures 102
  Change of Phloem and Periderm Structures 104

6 Modification of the Xylem and Phloem by Ecological Factors

Intra-Annual Density Fluctuations, Phenolic and Crystal Deposits 108
Intra-Annual Cell Collapse, Callous Tissue and Ducts 110
Interannual Variation of Latewood Zones 112
Long Term Variations: Sudden Growth Changes 113
Inter- and Intra-Annual Variations of the Phloem 114

7 Modification of Organs

Modification of Shoots:
  Long and Short Shoots 118
  Shedding Needles, Male and Female Flowers 121
  Thorns and Spines 122
  Vertical, Horizontal and Drooping Twigs 124
  Latent and Adventitious Shoots 126
The Lateral Modification of Stems 128

8 Anatomical Plasticity

Wood Structural Variability
  In Different Families 132
  In Different Growth Forms 134
  Under Different Site Conditions 136
Modification Caused by Different Shoot and Root Functions 140

9 Modifications Caused by Weather and Climate

Major Wood Anatomical Types in Different Climatic Regions 144
Modification of the Annual Tree-Ring Formation Caused By Seasonal Climatic Changes 148
Modification of the Annual Tree-Ring Formation Caused By Seasonal Climatic Changes:
The Genetic Component 150
Modification of the Xylem due to Intra-Seasonal Variations:
Ecological, Climatic and Individual Compontents 152

10 Modifications Caused by Extreme Events

Lack of Light 154
Severe Frost 158
Drought and Drainage 160
Defoliation by Insects 162
Defoliation Caused by Chemical Pollution and Nuclear Radiation 164
Crown Destruction due to Grazing 166
Crown Destruction Caused by Pruning and Pollarding 168
The Felling of Stems 170
Growing Together: Anastomosis 172
Crown, Stem and Site Destruction by Forest Fires 174
Crown and Stem Destruction by Parasites and Pathogens 178
Mechanical Stress on Stems due to Imbalance and Shock 180
Physiological Stress Caused by Stem Wounds 184

11 From Anatomical Features to Plant Structures

How do Woody Plants Get Old? 188
How Large Can Trees Get? 190
The Structural Diversity of Woody Plants 192
Protection Against Environmental Extremes
   Temperature Extremes 195
   Avoiding Shade 197
   Storage of Reserves in Seasonal Climates 198
Other Special Ecological Adaptations
   Herbivory and Ant Plants 199
   Mangroves and Flooding 200
   Mistletoes 201
   Phyllods, Phylloclades, Green Woody Stems 202

12 Decay of Dead Wood

Insects 204
Fungi 206
Carbonization 208
Petrification 210
Compression 212

13 Microscopical Preparation

Collection and Storing of Material and Preparation for Sectioning 216
Making Thin Sections 216
Preparation of Thin Sections for Permanent Slides 217
Observation and Photography 217

References 219
List of Species 221
Subject Index 227

Abbreviations

bpit	bordered pit
ca	cambium
cal	callus, parenchymatic cells
clu	cell lumen, cell lumina
co	cortex
cry	crystal
cu	cuticula
dv	density variation
ep	epidermis
en	endodermis
ew	earlywood
ewv	earlywood vessel
ewt	earlywood tracheid
ft	fiber tracheid
f	fiber
gr	growth ring
grb	growt ring boundary
lf	libriform fiber
lcw	lignified cell wall
lw	latewood
lwv	latewood vessel
lwt	latewood tracheid
pa	parenchyma
ph	phloem
phe	phellem
phg	phellogen
r	ray
rd	resin duct
rdil	ray dilatation
rtr	ray tracheid
sc	sclereid
si	sieve cell, sieve element
spit	simple pit
t	
tr	tracheid
ty	tylosis
ulcw	unlignified cell wall
v	vessel
vab	vascular bundle
xy	xylem

Technical Remarks

Colors in micro-photographs:
- In normal light with white background:
o Unstained slides: cell walls appear yellow and tannins brown
o Most slides are stained with Astrablue and Safranin.
- Red are lignified cell walls, tannins and sometimes nuclei.
- Blue are unlignified cell walls and sometimes nuclei.
Totally red slices are stained only with Safranin.

- In polarized light with black background:
o Cell walls with secondary walls appear light (birefringence). Starch and bordered pits reflect light in form of a Maltesian cross.
o Cell walls without secondary walls appear dark.

Magnifications are indicated in μm above a thin scale in each picture.

All thin sections were made with an old Reichert sliding microtome and the photographs with an Olympus microscope BX 51 with apochromatic objectives and the digital Olympus camera C 5050. For more details see chapter 13 p. 216.

This book describes the anatomical basics of wood technology (Table I.1) and dendrochronology. Long sequences, so-called chronologies and absolute dating are the magic words. The method of cross-dating gives each growth ring as it is shown in its microscopic structure in the present book, an absolute date, and a calendar date (I.2).

The *date* links the world. Absolute growth ring data allow the worldwide synchronization of cultural heritage. However, a basic requirement for this is the availability of regional centennial or millennial absolute chronologies. In fact, chronologies go back millennia to prehistoric times, and regionally many long chronologies exist in North and South America, in Europe, the Near East as well as in Siberia and in Eastern Asia, but these need to be synchronized. In addition, we would like to know more about the environmental conditions in the past. Again, tree rings offer this opportunity. Structural anomalies in single rings or in periods allow climate and weather reconstructions.

Figure I.2 gives an example for the synchronization of tree ring sequences. Using statistical tools, the largest similarity between overlapping pieces is being established. It emerges, that tree rings link different cultures even if they operated under very different conditions (I.3). The anatomical structure of growth rings expresses also the environmental situation during the growth period of the trees. An example is the so-called "Hungerjahr" in 1816 (I.4). In this year, maximum densities in conifers of northern latitudes and high altitudes are very low, i.e. the late wood formation was inhibited, although early season growth of the tree ring appeared quite normal. The missing late wood is an indicator for a cold summer in Europe, while the summer was dry and hot in western North America (I.5).

The book starts with the evolution of plant stems since the Palaeocoic. It explains the fundamental mechanisms of secondary growth, gives insights into the enormous structural variability triggered by mechanical and ecophysiological processes, and by biological defense mechanisms. The book carries anatomy to the level of morphology and explains how "cellular tools" are used in nature to produce the overwhelming variation in observed structures. We end with the description of structural changes during decomposition. In a final chapter we explain the methodology of studying wood anatomy.

Table I.1 Technical properties of wood species, steel and glass. Wood is light in comparision with other construction materials and has high compression, strength and bending values (Sell 1987, Fischer et al. 2002, and Richter, oral communication).

Species	Density g cm^{-3}	Compr. strength N mm^{-2}	Tensile strength N mm^{-2}	Bending strength N mm^{-2}
Fagus sylvatica, Beech	0.65	60	120	100
Quercus robur, Oak	0.65	65	100	95
Diospyrus crassiflora, Ebony	1.05	80	ca. 200	189
Standard concrete	2.5	30-50	2-4	3-8
Construction steel	7.8	150-235	360-510	170-330
Construction glass	2.5	700-900	30-90	45

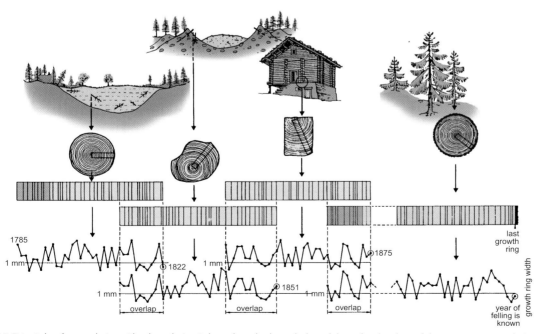

I.2 Principle of cross dating. Absolute dating is based on the knowledge of the calendar date of the youngest growth ring (below the cambium) of a living tree. The occurrence of irregularly formed rings enables the chronological bridging of absolutely dated with not dated growth ring sequences.

I.3 Contemporaneous buildings of very different cultures of the 13th century with wooden construction elements: the Lom stave church in Norway; a cliff dwelling in the arid climate of the American Southwest (Betatakin, Anazazi culture) contains conifer beams in walls; a wooden temple in Hakuro, Japan.

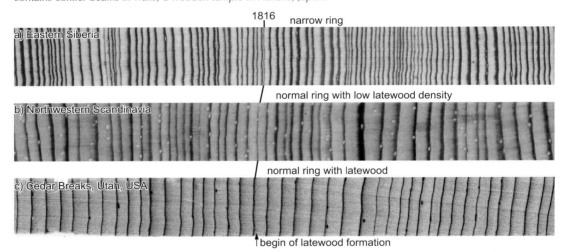

I.4 Conifer ring sequences including the year 1816 from different climatic situations at different geographic positions. The year 1816 is known as „the year without summer", or „the famine year". Trees tell a much more differentiated story e.g. a) in Eastern Siberia, the ring is small, which indicates poor conditions for growth in spring and summer b) the latewood density is reduced in pines of Northwestern Scandinavia, which indicates a cold summer c) growth was just normal in the subalpine zone of the Rocky Mountains in Utah, USA. The tree rings prove and document facts which are otherwise only known from history. Density assessment by X-ray photographs from cores were used for the construction of hemispherical dendroclimatological maps.

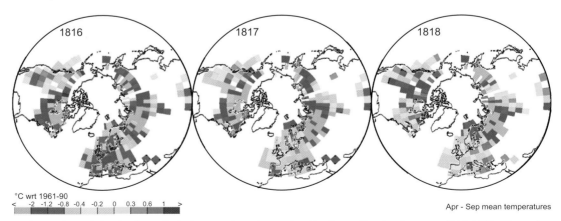

I.5 Hemispherical dendroclimatological maps of maximum densities in conifers for the years 1816-1818. Blue colors indicate cold summers, and red colors indicate warm summers. It is obvious that most part of the hemisphere had an extremely cold summer in 1816; in contrast, the summer was warm in the American West. Therefore the famine year was not a global event (Briffa et al. 2002).

CHAPTER 1

THE EVOLUTION OF PLANT STEMS IN THE EARTH'S HISTORY

Evolution created a large diversity of different types of stem constructions. This chapter follows the principal steps through geological times and illustrates them by means of "living fossils".

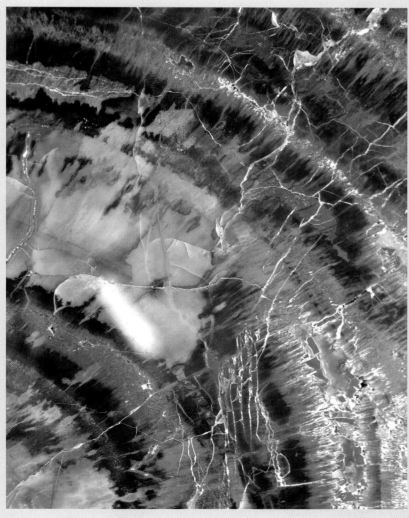

Cross section of a petrified stem of Araucaria. Triasssic, 200 million years before present. Arizona, USA.

The Landscape in the Paleozoic

Plant life developed in the water, and this process took over three billion years. Only relatively recently - speaking in terms of the Earth's history - when the atmosphere's oxygen content had reached suitably high levels, terrestrial environments were colonized by plants. The so-called "Step on land" took place about 450 million years ago on permanently humid sites in coastal environments. At that time, the Earth generally had a warm climate. Only some spores, fragments of cuticles and epidermis are early witnesses from this period.

Evolution and specialization was not a continuous process. Major astronomic and volcanic events caused repeated extinctions of the dominant life forms. However, life continued; extinct species were replaced by species which were genetically broader. The broad spectrum of allels might not even have been beneficial before the extinction event, but became a crucial advantage after the event. These species survived extreme conditions and exhibited an accelerated speciation when re-occupying the devastated habitats of the Earth. All extinction events resulted in a change of the dominant flora and fauna. Thus, the evolution of stems, as discussed in the following, is synchronized by these external forces.

The first extinction at the end of the upper Cambrian hit the Trilobites, caused the reduction of Stromatolithes and accelerated the specialization of vertebrates. A minor extinction in the late Devonian promoted the spread of tree-like ferns (Pteridophytes). The most devastating extinction event happened at the end of the Permian (1.1, 1.2). It eliminated the Lycopod and Horsetail forests and initiated the spread of the Gymnosperms. The event at the end of the Cretaceous eliminated the dinosaurs and most Lycopsidae. It created space for the evolution of the Angiosperms. Figure 1.2 does not show the so-called present extinction caused by the human population.

The „Step on land" taken by the plants was in reality a fundamental step that has determined the aspect of the continents to the present day. The attraction for life on land was the higher availability of CO_2 in the atmosphere. However, the constraints were a lower supply of water, and the loss of water as support for the plant structure. This resulted in changes of plant physiology, morphology and anatomy. It is evident that these changes did not take place in just one event.

The earliest plant fossils of stems are known from the Silurian and Devonian (439-409 million years ago; 1.3). These belonged to the evolutionary line of club mosses (Lycopodiopsida). At that time, only a small part of the large continent Gondwana was located at the equator. This is most likely the region where plants started to grow outside the water. Over the following 120 million years, the continent drifted northwards (1.1). The large continental area contained rather different climatic conditions and this in turn enhanced a diversification of the terrestrial flora and the formation of real ecosystems. Forests with tall trees developed during the Carboniferous (363-290 million years ago). These remained near the equator (1.4). Today, coniferous forests are dominant in eastern North America, Europe and Siberia.

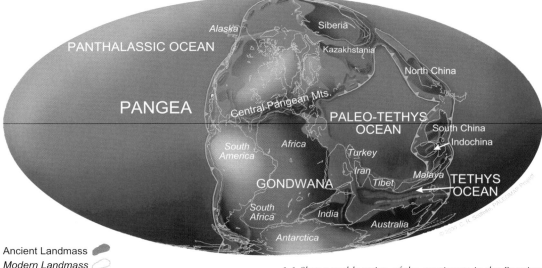

Ancient Landmass
Modern Landmass
Subduction Zone (triangles point in the direction of the subduction)
Sea Floor Spreading Ridge

1.1 Shape and location of the continents in the Permian, 255 million years ago. The land area had shifted towards the equator. A volcanic event caused the extinction of the Lycopod and Horsetail forests (Scotese 1997).

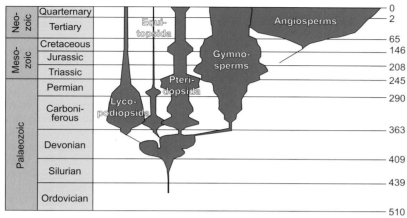

1.2 Relative species diversity of the most important terrestrial plant groups since the beginning of the Ordovician (after Strasburger et al. 2002).

1.3 Landscape reconstruction in the lower Devonian, 409 million years ago. Big „stems" of aquatic algae had drifted to the sea shore. Tiny lycopods grew on moist lake shores. Algae: (1) Prototaxis sp.; Psilotales: (2) Taeiniocrada decheniana, (3) Zosterophyllum renanum, (4) Sciadophyton steinmannii; Lycopsida: (5) Drepanophycus spinaeformis, (6) Protolepidodendron wahnbachense (Scharschmidt 1968).

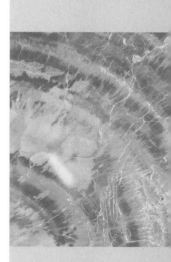

1.4 Landscape reconstruction in the upper Carboniferous period, 300 million years ago. On moist sites, real forest had become established, forming a tropical-type ecosystem with plants of different heights, adapted to diverse ecological conditions. (1) Lepidodendron sp., (2) Sigillaria sp., (3) Stylocalamites, (4) Calamitina, (5) Crucicalamites, (6) Sphenophyllum cuneifolium, (7) Megaphyton, (8) Stuaropteris (Scharschmidt 1968).

Plant Body of Vascular Plants (Cormus)

The principal differentiation of plants advanced when they colonized the land. The shoot, which grows mainly above the ground, developed a main axis, supporting branches and leaves. Below ground, the roots are characterized by high ramification (1.5).

The anatomical stem structure of early land plants appears to be very simple, but in reality it includes all elements of every contemporary terrestrial plant. This becomes evident from fossilized plants (1.6, 1.7). An epidermis with a cuticle surrounds parenchyma tissue and a simple vascular bundle (1.8, 1.9). The gas exchange with the atmosphere (CO_2 and water vapor) occured through stomata. *Rhynia* and *Asteroxylon* already had a fungal symbiosis, a mycorrhiza (Pierozynski and Malloch 1975).

In the Devonian, plants were already differentiated into tissue which protected the plant against drought, tissue which was capable to store organic matter, and tissue which conducted water and assimilated and gave stability. Other major differentiation processes already existed: cell division, cell enlargement, cell differentiation and cell wall differentiation, lignification and cell wall growth (see p. 34). Also, all actual physiological processes, such as photosynthesis and dissimilation, had already evolved.

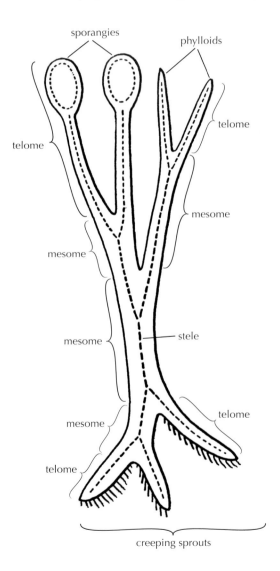

1.5 Conceptual drawing of the first land plant (Zimmermann 1959).

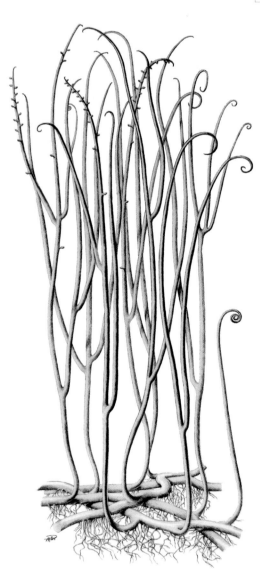

1.6 Reconstruction of a plant from the lower Devonian, *Rhynia major*, which was characterized by a main stem, a bifurcate branch system, rhizome and small fine roots (Schweizer 1990).

1 The Evolution of Plant Stems in the Earth's History

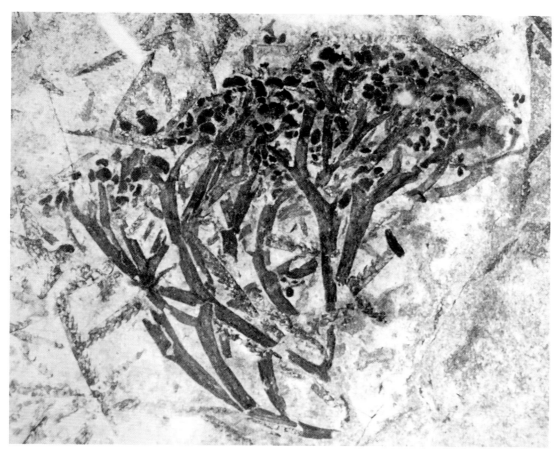

1.7 One of the first land plants. Fossilized specimen of Cooksonia devonica from the lower Devonian, characterized by a dichotomy of the telom (Schweizer 1990).

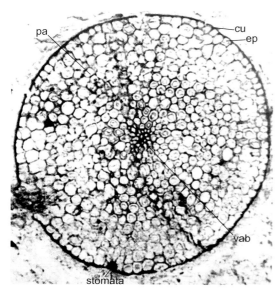

1.8 Anatomical structure of the first land plant, Rhynia major. Stem cross-section. Characteristic is the epidermis with stomata, parenchyma and a central vascular bundle (Zimmermann 1959).

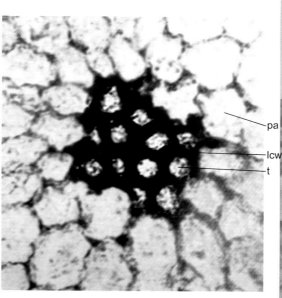

1.9 Cross-section of a vascular bundle in the stem of Rhynia. Characteristic is the xylem, which consists of thick-walled tracheids (Zimmermann 1959).

The Evolution of a Stabilizaton System

The cells of present-day aquatic plants are thin-walled and generally do not have a secondary wall, as there is no need to support the tissue's body weight. However, there were early Devonian waterplants which had sufficient structural tissue to hold the sporangia above the water surface for sporal distribution (1.10). Thus, there were pre-adaptations in water plants which became important for terrestrial plants. Nevertheless, most of the first small land plants - we only know them from their spores - remained upright mainly by high osmotic pressure.

The earliest terrestrial plants known had stable cell walls with a high lignin content. This particular „invention" allowed the plants to form a vertical stem. Furthermore, the earliest terrestrial plants were already able to produce eccentric stems, i.e. they could direct photosynthetic products to where they were needed in order to maintain stability (Jung 1986; 1.11, 1.12).

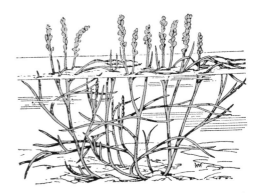

1.10 Fossil Devonian water plant with sporophylls above the water surface. Zosterophyllum renanum (Kräusel and Weyland 1926).

The Evolution of a Water-conducting system

The earliest true terrestrial plants already possessed a well-differentiated water-conducting system. This ancient system transported water from one live parenchyma cell to another, and this remains an important transport mechanism until today. In modern plants, for long distances water flows through the dead tracheids and vessels in the xylem which connects the living roots in the soil with all live cells above the ground. The photosynthetic assimilates flow through live sieve tube cells in the phloem to all non-photosynthesizing live cells.

The Evolution of a Protection System

The great water pressure deficit between air and cell wall (0-40 hPa) forced the plants to develop a boundary system between the dry air and a water-saturated plant body to protect them against desiccation. The earliest terrestrial plants already had an epidermis with a cuticle and air-permeable stomatal pores.

For plants that do not live in the water, there is a great risk that their bodies may become desiccated. For this reason, plants had to develop defence mechanisms against drying out. The early terrestrial plants had cell walls with pits, which allowed them to isolate infected or dead cells from healthy cells by barriers. Furthermore, the plants probably developed toxic phenols which became the basis for lignin synthesis.

1.11 Fossil terrestrial plant Hyenia from the lower Devonian, in the German Rhine valley (Schweizer 1990). It shows a differentiation into stems and leaves.

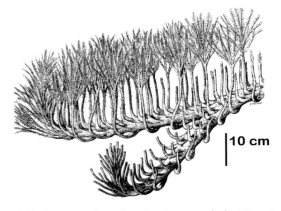

1.12 Reconstruction of the fossil terrestrial plant Hyenia from the lower Devonian, in the Rhine valley.

THE CONTEMPORARY FOSSIL *PSILOTUM NUDUM*?

The morphological and anatomical structure of the contemporary pantropical genus *Psilotum* (1.14) is very similar to that of the Devonian *Rhynia major*. The species has a large cortex (1.13), a vascular bundle with living cells (1.16) and an epidermis with stomata (1.15). According to Judd *et al.* (2002), however, DNA studies indicate a relationship to ferns. Despite this controversy, the plant demonstrates what the first land plant might have looked like.

Psilotum nudum (Whisk Fern) has a root-like system that anchors the plant, and a photosynthesizing stem.

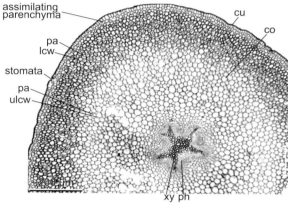

1.13 Stem anatomy of Psilotum nudum. The central, star-shaped vascular bundle is surrounded by a large body of parenchyma cells. Stability is obtained by a tube of thick-walled, lignified sclerenchymatic cells. A two-cell layer of thin-walled cells contains chloroplasts. An epidermis with a thick cuticle protects the plant against desiccation. Air can be exchanged through a few stomata.

1.14 Morphology of Psilotum nudum. Note the erect, well-branched shoots above the ground, the scale-like leaves and the sessile yellow sporangia. Hinchinbrook Island, Queensland, Australia.

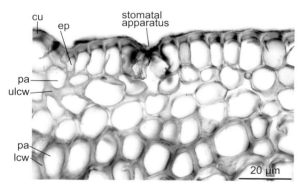

1.15 Peripheral cell anatomy. The stem is sealed by a thick cuticle, and simply constructed stomata allow an exchange of carbon dioxide and oxygen. Photosynthesis takes place in chloroplasts, which are - here destroyed by preparation - located in thin-walled parenchyma cells below the epidermis.

*1.16
Left: Longitudinal section through the xylem of the vascular bundle; characteristic are the tracheids with pits with slit-like apertures.*

Right: Radial section through the phloem of the vascular bundle; one sieve cell contains a nucleus, the other shows the sieve areas at the radial side.

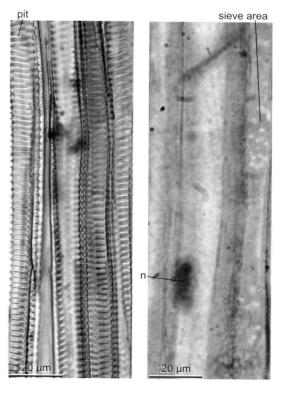

Diversification of Plants Containing Tracheids
Lycopods

The Carboniferous is characterized by a great diversification of spore-producing species containing tracheids (Tracheophyta). One evolutionary branch led to the formation of the lycopods (Lycopodiopsida). Large parts of the swampy Carboniferous forest consisted of numerous species from the *Lepidodendron* and *Sigillaria* (1.17) genera, but these became extinct in the Permian extinction event (1.1).

The Lycopod stems were highly developed and had a cambium with secondary growth. However, in contrast to the present trees, the Lycopods produced „bark stems" because the cambium produced mainly bark (1.18); there was very little xylem (1.19). This structure restricted water transport, and most likely it was water stress which confined these plants to live in swamps. They did not occupy dry terrestrial land. From this group, only the Club Moss (Lycopodiales) and Moss Fern (Selagininellales) have survived to the present day, though all of them have not obtained the ability to produce secondary xylem.

Our knowledge of plant development in this stage is restricted to wet habitats which were suitable for fossilification.

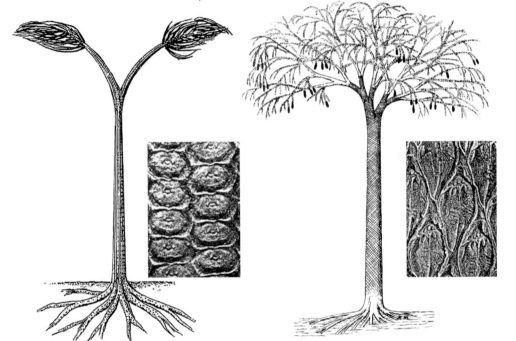

1.17 Sigillaria (left) and Lepidodendron (right) trees from the Carboniferous with dichotomously branching root and crown systems and the well-preserved leaf scars (small pictures; Scharschmidt 1968).

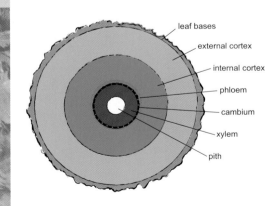

1.18 Lepidodendron "bark stem". Characteristic is the large cortex and the small xylem (after Scharschmidt 1968).

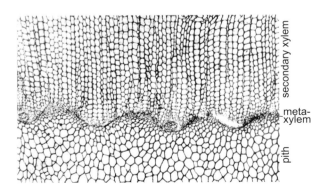

1.19 Sigillaria saullii xylem. The pith is separated from the secondary xylem by a zone containing the protoxylem and metaxylem. The tracheids are characteristic of this plant (Zimmermann 1959).

DIVERSIFICATION OF PLANTS CONTAINING TRACHEIDS
HORSETAILS

One evolutionary branch led to the formation of horsetails (Equisetopsida; 1.21, Judd et al. 2002). During the Carboniferous, many species of the genus *Calamites* formed 30 m high trees with stems containing big piths (1.20), but the trees were unable to produce a secondary xylem.

The genus *Equisetum* survived mainly as herbaceous plant growing on moist sites in temperate and boreal zones (1.22). Characteristic is the tissue with air ducts (1.23).

1.20 A fossil Calamites. The Calamites formed „pith stems" (Zimmermann 1959). The large pith and the xylem that consists of tracheids are characteristic of this plant.

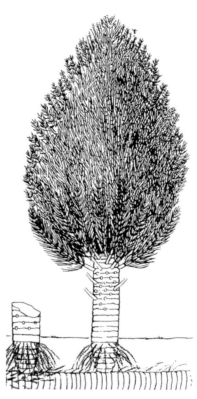

1.21 Reconstruction of a fossil, extinct, tree-like horsetail, Calamites sp. (Judd et al. 2002).

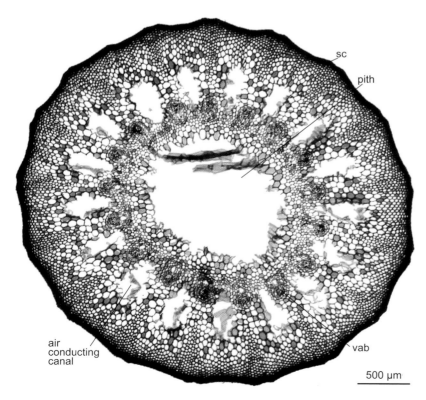

1.22 Fertile shoots of Equisetum maximum (Giant Horsetail).

1.23 Microscopic section of a modern *Equisetum hiemale* (Rough Horsetail) stem. The closed vascular bundles around the pith, the big air conducting canals and the sclerenchymatic peripheral zone are typical for horsetails.

Diversification of Plants Containing Tracheids
Fossil and Contemporary Ferns

Another branch in evolution led from the Psilophytopsidae to the formation of the ferns (1.24; Judd et al. 2002). Fern trees were most abundant in the Carboniferous. Also, ferns did not develop a secondary xylem. Stem stability was assured by a thick mantle of roots (1.26) or leaf bases (1.25). Secondary growth was absent (1.27). The anatomy of the contemporary genus *Osmunda* is still the same as in fossile ferns (1.28, 1.29).

1.24 Reconstruction of the fossil tree-like fern Psaronius (Zimmermann 1959).

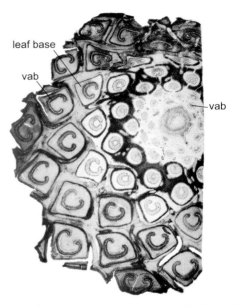

1.25 Stem of the tree fern Thamnopteris schleichedalii. The central stem is surrounded by leaf bases (Zimmermann 1959).

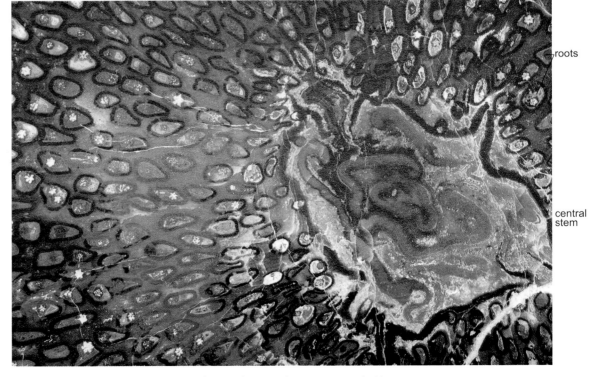

1.26 Stem base of the fern Psaronius brasiliensis. The central stem is surrounded by roots (Jung 1986).

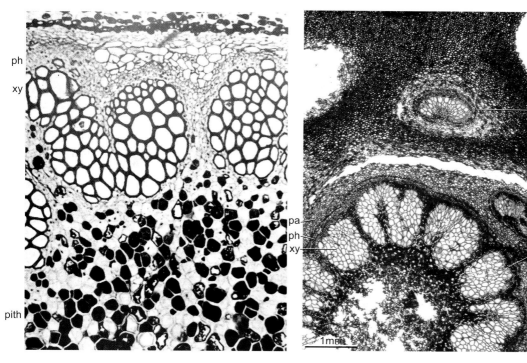

1.27 Vascular bundles surround the big pith of a fossil fern stem (Osmundales). The xylem with big, thick-walled tracheids and the phloem with thin-walled cells at the periphery are characteristic. A cambium is absent (Zimmermann 1959).

1.28 Vascular bundles of a contemporary Royal Fern (Osmunda regalis) with the xylem surrounding the pith. There are extern leaf vascular bundles. Ticino, Switzerland.

1.29 Root stock of the Royal Fern (Osmunda regalis), Ticino, Switzerland. The principal construction is very similar to that of the Carboniferous Psaronius. Small right: from Aeschimann et al. 2004.

1 The Evolution of Plant Stems in the Earth's History

Diversification of Plants Containing Tracheids
Contemporary Ferns

A large number of fern genera and species with a variety of growth forms, such as trees, lianas, epiphytes and herbs (1.30, 1.32, 1.34), have survived to the present day. Ferns adapted to most of the present environmental conditions, e.g. fern trees grow in tropical and subtropical zones, and small herbs are able to survive in arid, arctic and alpine regions. Characteristic for fern stems is the absence of secondary growth (1.31-1.36) and the presence of centric vascular bundles (1.36). Stability is reached by thick sclerenchymatic belts.

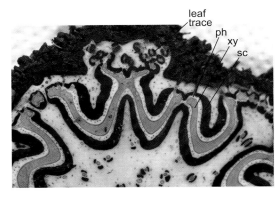

1.30 The Tasmanian Tree Fern (Dicksonia antarctica) in Australia. The uniform stem with a fan-shaped bundle of leaves is typical of this plant. Small: Dicksonia antarctica canopy, Westland, South Island of New Zealand.

1.31 The stem of the treefern Dicksonia antarctica. The big pith is surrounded by a wavy vascular bundle. In its center there is a light-brown band of sclerenchymatic cells. The stem is surrounded by leaf bases.

1.32 Macroscopic cross-section of the stem base of the Bracken Fern (Pteridium aquilinum). The German name is „Eagle Fern" (Adlerfarn) because the vascular bundles form a „double eagle".

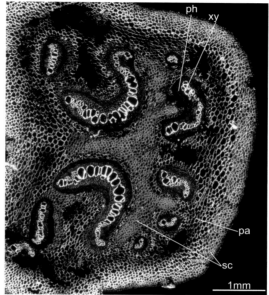

1.33 Microscopic cross-section of the stem of Bracken Fern (Pteridium aquilinum). There is a star-shaped sclerenchymatic column in the center surrounded by vascular bundles. Big, white tracheids form the xylem, and the sieve tubes are found on both side of the vascular bundle. The stem is surrounded by a dense mantle of sclerenchymatic cells (polarized light).

1.35 Broad Buckler Fern (Dryopteris austriaca) in a montane Picea abies forest, Rhoen, Germany.

1.34 Base of the herb-like Male Fern (Dryopteris filix-mas). The two very small „stems" are surrounded by leaf traces. This construction is comparable with the fossil fern stem shown in 1.25.

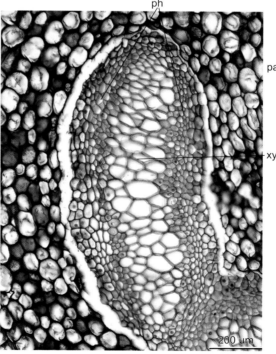

1.36 Vascular bundle of the Green Spleenwort (Asplenium viride). The bundle is embedded in parenchymatous tissue. There is phloem on both sides of the xylem that contains thick-walled tracheids.

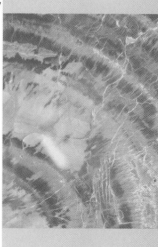

1 The Evolution of Plant Stems in the Earth's History

Trees Grow Taller and Bigger

A major step in evolution towards woody stems occurred during the late Devonian: secondary growth developed, which allowed the formation of a substantial trunk suitable to conduct sufficient water for the canopy (1.37, 1.38). Most of our contemporary trees evolved from this evolutionary step taken by the seed plants.

Secondary growth became possible by re-organization of the vascular bundle from closed bundles towards open vascular bundles. The inner xylem is separated from the outer phloem and bark by a long-lived meristem.

The first seed plants appeared 360 million years ago in the early Carboniferous (Judd *et al.* 2002). Many of these became extinct, but the major gymnosperm tribes present on Earth today evolved from these species. The „invention" of the stem with a cambium, xylem, phloem and periderm was the basis for invading the dry habitats of the continents. Gymnosperms were co-dominant in the Carboniferous lycopod forest. All gymnosperms, such as the fern-like *Archaeopteris* (1.37) and the conifer-like Cordaopsidae (1.38), were able to produce secondary growth. From a morphological or anatomical point of view they resemble modern conifers (1.39-1.41). The structure of both is similar to that of modern *Araucaria*.

The *Archaeopteris* and the Cordaopsidae became extinct at the end of the Palaeozoic, but their capacity to produce secondary growth survived in conifers and angiosperms.

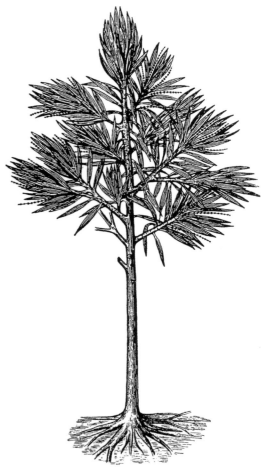

1.37 Reconstruction of the Devonian fern-like tree Archaeopteris. The conical stem produced secondary growth. Hence, the stems could grow to 1.5 m diameter and reach a height of over 15 m (Schweizer 1990).

1.38 Reconstruction of a Cordaites tree, a Carboniferous gymnosperm. The long leaves and distinct stem with secondary growth are characteristic of this plant (Zimmermann 1959).

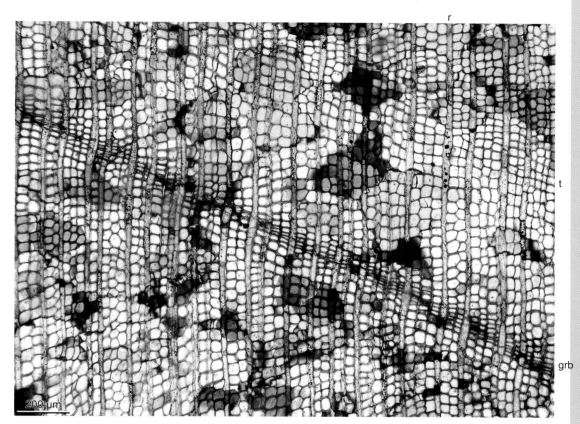

1.39 Microscopic transversal section of the petrified wood of Dadoxylon sp., a species of the extinct Cordaites family. Characteristic is the uniform conifer-like anatomy. The wood consists of tracheids and parenchyma. Its colouring is caused by the particular crystallization of the different minerals (polarized light). Boulder in Quaternary sediments, Jura Mts., Switzerland.

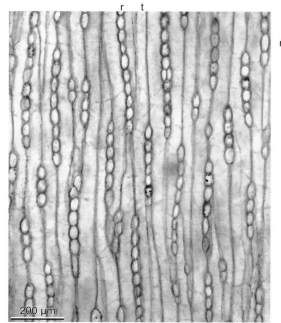

1.40 Microscopic tangential section of Dadoxylon sp. wood. The rays are mostly uniseriate.

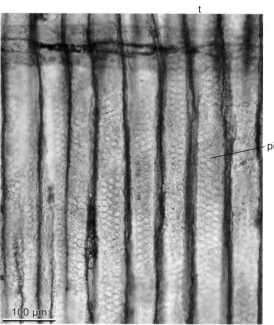

1.41 Microscopic radial section of the Dadoxylon sp. wood. Note the characteristic two or three rows of pits on the tracheids (araucariod pitting).

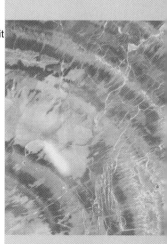

1 THE EVOLUTION OF PLANT STEMS IN THE EARTH'S HISTORY

Successful Seed Plants with Naked Seeds
Ginkgos and Cycads

The Permian was characterized by a very variable climate with very dry and wet periods. Under these conditions many old tribes became extinct, but a large number of new tribes were able to develop: Ginkgos (1.42), Cycads (1.43, 1.44), Gnetophytes and conifers. All of them still exist today; the conifers still dominate entire regions, whereas all the other genera only survived as scattered relics. The whole group is termed gymnosperms, with reference to their naked seeds, as opposed to angiosperms, in which the seeds are enclosed inside a carpel.

Only one species of *Ginkgo* has survived, and today it is mainly found around temples in China. A great diversity of Ginkgos existed in the Mesozoic (245-146 million years ago). They are characterized by fan-shaped, bifurcate leaves and conifer-like wood (1.45) with secondary growth.

Cycads were most abundant and diverse during the Mesozoic. Approximately 130 species have survived, mostly in tropical and subtropical regions. Secondary growth is limited (1.46) or absent (1.47).

1.42 Ginkgo biloba leaves and male flowers. The fan-shaped leaf venation is characteristic of this tree. Botanical Garden, Würzburg, Germany.

1.43 Cycas armstrongii. The palm-like growth form is typical.

1.44 Macrozamia communis. A very short, bulb-like stem is characteristic of this plant. It lives in symbiosis with blue-green algae, encapsulated in coral-like specialized roots for N_2-fixation (small picture). West Australia.

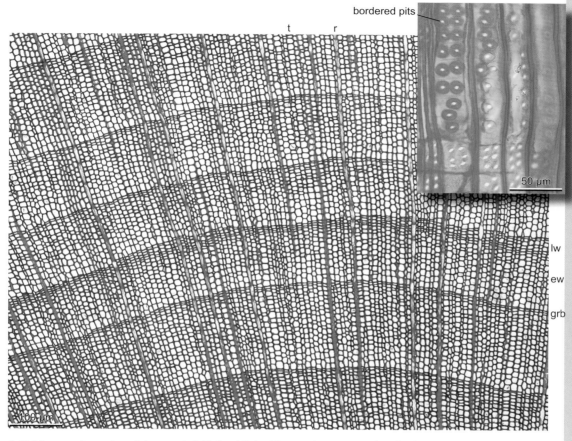

1.45 Microscopic section of the wood of Ginkgo biloba. The simple anatomy of tracheids is characteristic of conifers. Top right: Radial section with bordered pits.

1.46 Cross-section of a Cycas armstrongii stem. The big pith is surrounded by a two-layered xylem and a phloem produced by successive cambia. The big cortex contains slime ducts.

1.47 Cross-section of a bulb-like Macrozamia communis stem. This very small stem is surrounded by live leave bases. It protects the plant from fire.

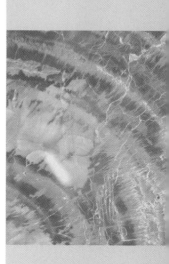

1 THE EVOLUTION OF PLANT STEMS IN THE EARTH'S HISTORY

Successful Seed Plants with Naked Seeds
Gnetophytes (*Ephedra*, *Gnetum* and *Welwitschia*)

The Gnetophytes are a morphologically very heterogeneous group: most *Gnetum* species are tree-like with broad leaves and grow in tropical forests (1.48, 1.49).

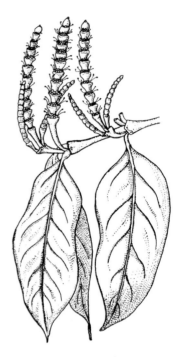

1.48 The Gnetophyte Gnetum gnemon. The leaves are similar to those of dicotyledons (left photo: Lange, Papua New Guinea; right: from Strasburger et al. 2002).

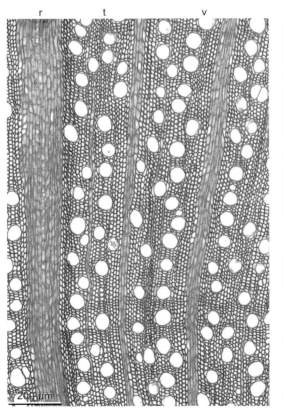

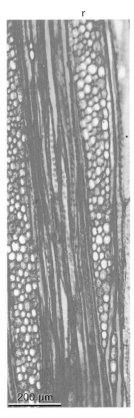

1.49 Microscopic sections of Gnetum gnemon. Left: Cross section - the wood structure resembles that of dicotyledons. Middle: Radial section - the conifer-like bordered pits in the tracheids demonstrate taxonomic relationships to conifers. Right: Tangential section - a distinction to conifers are the large rays.

Only one *Welwitischia* species survives in the African Namib Desert (1.50, 1.51). Common to all three genera is wood similar to that of angiosperms, which contains large rays and vessels (1.53).

1.50 The Gnetophyte Welwitschia mirabilis with two above-ground leaves and a beet-like stem below the ground. The two leaves exhibit a meristem at the base of the leaves. Thus, the leaf grows at the base and dies off at its tip. The plant persists for up to 3000 years (Judd et al. 2002).

1.51 Old Welwitschia mirabilis in the Namib Desert, Swakop river, Namibia. Welwitschia has male and female plants; this one is female (photo: Poschlod).

All *Ephedra* species are similar to angiosperm shrubs; they have scale-shaped leaves and grow on dry sites (1.52, 1.53).

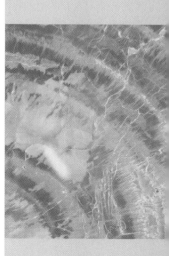

1.52 The Gnetophyte Ephedra sp. (Joint Pine) in the arid region of the North American west.

1.53 Microscopic cross section of Ephedra major wood. Typically, the structure resembles that of the dicotyledons. The xylem consists of tracheids, parenchyma and vessels.

1 THE EVOLUTION OF PLANT STEMS IN THE EARTH'S HISTORY

21

The Most Successful Seed Plants with Naked Seeds
Conifers

The first true conifers, the Voltziales, emerged in the late Palaeozoic (Carboniferous) and existed until the Jurassic (1.54). In comparison with modern plants, they are most closely related to the Pinaceae, which evolved in the Jurassic.

Five major groups of conifers survived all climatic changes: Pinaceae, Araucariaceae, Cupressaceae, Podocarpaceae and Taxaceae.

Araucaria species have survived in relic forests of the southern hemisphere, whereas the Cupressaceae form a large group of species mainly in dry climates. Some hundreds of species of Pinaceae form the boreal zone of the northern hemisphere (1.55). Major representatives of this family are the genera *Pinus*, *Picea*, *Abies* and *Larix*. Many *Podocarpus* species are found in tropical forests, while *Taxus* is an understory species of temperate broad-leaved forests. *Araucaria* forests were most abundant during the Mesozoic, 200 million years ago, forming a forest belt around the equator (1.56). Cupressaceae, such as the genera *Juniperus* and *Sequoia*, grow as small shrubs up to tall trees (1.57).

Conifer xylem is surprisingly homogeneous and consists only of tracheids and parenchyma (1.58-1.60).

1.54 Reconstruction of a Permian conifer, Lebachia piniformis, which belongs to the order of Voltziales (Zimmermann 1959).

1.55 Black Spruce forest in the Canadian boreal forest. Picea mariana represents the Pinaceae family.

1.56 Araucaria forest at the timberline in the Chilean Andes. The Monkey-puzzle Tree (Araucaria araucana) represents the Araucariaceae family.

1.57 Juniper forest at the timberline in the Kirgisian Tien-Shan Mountains. Juniperus turkestanica represents the Cupressaceae family.

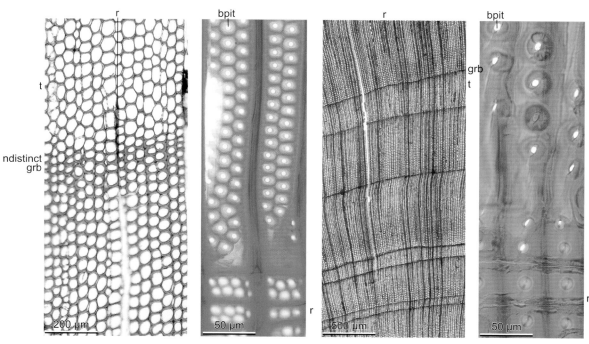

1.58 Microscopic section of Parana Pine wood (Araucaria angustifolia), Chile. Typically, the resin ducts are missing (left), but there are several rows of bordered pits in alternating position on tracheids (right).

1.59 Microscopic section of Alligator Juniper wood (Juniperus depeana), Arizona, USA. Typically, resin ducts are missing (left). Bordered pits on tracheids are uniseriate (right).

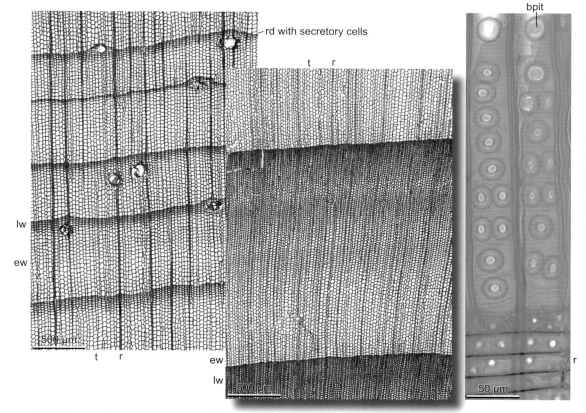

1.60 Microscopic transversal section of Scots Pine (Pinus sylvestris, left) with resin ducts, and Silver Fir (Abies alba, middle) without resin ducts. Bordered pits on tracheids are uniseriate or biseriate in opposite position (radial section right).

Successful Plants with Seeds Enclosed in a Carpel
Angiospermae

The first known ecosystem consisting of angiosperms dates to the lower Cretaceous, ca. 140 million years ago (Zhou *et al.* 2003), but the major worldwide evolutionary change took place in the upper Cretaceous, ca. 100 million years ago. The actual morphological and anatomical diversity of angiosperms is enormous. Approximately 250,000 species are known. These have many features in common: in the flowers, the sieve tubes and in the companion cells of the phloem. At the base of the phylogenetic tree, there are some families of the sub-class Magnoliidae. One of the most primitive species is *Amborella trichopoda* (Amborellaceae) from New Caledonia (Endress and Igersheim 2000; 1.61, 1.62). Taxonomically close to this are the Winterales, the Piperales (1.63, 1.64) and the Magnoliales (1.65, 1.66).

Despite morphological similarities in the flowers, the xylem's anatomy is very heterogeneous. The anatomy of cell types and tissue even of these early angiosperms is highly variable: some species have vessels with scalariform or open perforations (1.66); some have just tracheids, but no vessels, such as *Amborella* (Carlquist and Schneider 2001; 1.62), whilst others are semi-ring-porous (1.64).

1.61 Amborella trichopoda, a representative of the Amborellales. A New Caledonian shrub (photo: Endress and Igersheim 2000).

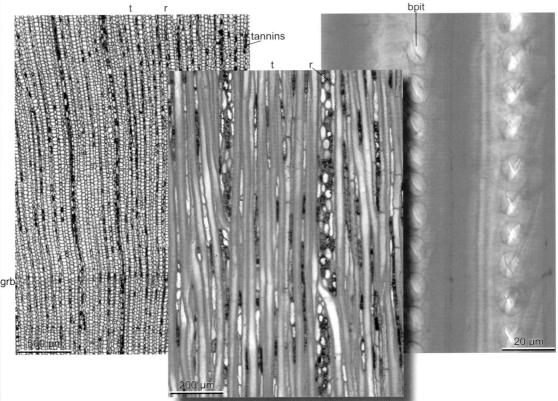

1.62 Amborella trichocarpa wood. The wood consists of tracheids and the missing vessels are characteristic of this plant (left). The growth ring boundary in this tropical tree is an expression of a period of dormancy. Relationships to Angiospermae are indicated by large rays (middle). The conifer-like bordered pits on tracheids show relationships to Gymnospermae (right; slides: Carlquist).

1.63 Birthwort (Aristolochia clematitis), a representative of the Piperales. A Central European, hemicryptophytic herb.

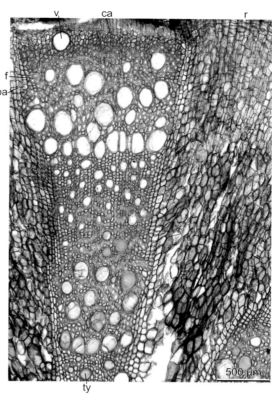

1.64 Aristolochia clematitis xylem. Semi-ring porosity and very large rays are typical.

1.65 Magnolia stellata, a representative of the Magnoliale, an East Asian tree. Jena, Germany.

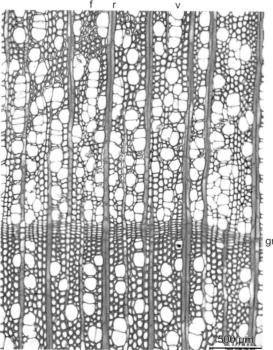

1.66 Magnolia virginianum wood. The diffuse-porous distribution of vessels is typical.

Systematic of Plant Life

The phylogenetic tree of plant evolution contains surprises which were not apparent from morphology (1.67). Starting with the Tracheophyta, the Lycopodiophyta remain as the group closest to the first land plants while *Psilotum* emerged as a degenerated fern. The Angiospermae separated from the ferns and the horsetails at a very early stage in parallel to the Gymnospermae where *Ginkgo* and *Cycas* had a separate evolution from the Pinophyta and the Gnetophyta. Within the Angiosperms, the Amborellaceae, an endemic family of French New Caledonia (NE of Australia), turned out to be closest to the early evolution of the Spermatophyta. The Magnoliids appear to be more closely related to the Monocots than to the rest of the Eudicots. These developed, via a number of unique families and orders, into two major groups, the Rosids and the Asterids. Thus, the phylogenetic tree has simplified the classification of the overwhelming diversity of morphological traits. It is the aim of this book to put one aspect of the morphological variation into the context of plant evolution, namely wood anatomy.

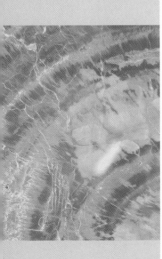

1.67 The evolution of plant life according to DNA sequences of the 18S rDNA, rbcL- and atpB-genes (Wink 2005, Wink, personal communication). The DNA sequences provide a quantitative basis for estimating the similarity between taxa, and the construction of the phylogenetic tree of evolution which is the basis for a modern taxonomy of plants. It will remain a major task for the future to relate the evolution of morphological traits to this DNA-based phylogeny. In the Figure, orders from which taxa are described in this book are printed in green, and an attempt is made to investigate the evolution of wood in this context.

Chapter 2

The Structure of the Plant body

The following chapter illustrates the principal cell elements and the construction of the plant body.

Alpenrose (Rhododendron ferrugineum).

Life Forms in Different Vegetation Zones

Three major factors determined the diversity of terrestrial plant life: the present geographical position as surrogate for climate, the history of climate, and horizontal and vertical tectonic movements at geological time scales. The result is a planet with very diverse marine and terrestrial systems (2.1) The climatic and edaphic conditions led to the development of taxonomically, morphologically and anatomically very diverse plant stems. □

Open shrubland/tundra: Arctic vegetation with Dryas integrifolia hummocks on Banks Island, Canada, 72°N

Deciduous broadleaf forest: Temperate deciduous forest with Beech (Fagus sylvatica) in Thuringia, Germany, 51°N

Evergreen needleleaf forest: Boreal forest with Black Spruce (Picea mariana), Great Slave Lake, Canada, 62°N

Closed shrubland: Chaparral near Riverside, California, USA, 33°N

Evergreen broadleaf forest: Rainforest, Costa Rica, Central America, 18°N

Steppe: Steppe vegetation of the southern hemisphere with Feather Grass (Stipa sp.) and Mulinum spinosum shrubs, near Escuel, Argentina, 43°S

Savannah: Subtropical savannah with Blue Thorn (Acacia erubescens) and numerous perennial C4 grass species, Uis, Namibia, 22°S

2.1 Simplified map of potential vegetation without human influence. This map was derived from the overlay of three satellite-based land cover datasets. Anthropogenic landuse was filtered out (Jung et al. 2006).

Principal Growth Forms of Stems

The structure of the Devonian plant body - root, stem, branch - has been successful until the present day. Over time, this basic structure has been altered in many ways, but two main types remained:

a) The palm-like growth form: at the top of a stem, there is a fan of leaves and flowers. The stem has closed bundles. Secondary growth is the exception. Most monocotyledonous plants belong to this form (2.2).
b) The tree-like growth form: the stem supports a branched crown. The stem has open bundles with secondary growth. Conifers and dicotyledonous plants belong to this form (2.3, 2.4).

The two main growth forms above reflect the presence (pyramidal growth of stems) and absence (cylindrical growth of palms) of secondary growth.

Characteristic for palms stems are the large vessels which are arranged in closed bundles (cable stem) (2.5), characteristic for temperate and boreal conifers is the xylem showing rings with distinct earlywood-latewood transitions (2.6), and characteristic for broad-leave trees stems is the great diversity of wood anatomy, here presented by a ring porous wood (2.7).

2.2 Monocotyledonous plant. Date Palm (Phoenix dactylifera). On top of a single, slim stem of uniform thickness is a shock of big leaves. Maroc.

2.3 Conifer. Norway Spruce (Picea abies). Many conifers, such as fir and spruce, have branches arranged in a radial direction around a central, conic stem. Alps, Switzerland.

2.4 Dicotyledonous tree. Salmon Gum (Eucalyptus salmonophloia). Most dicotyledonous trees have a well-developed branch system, which forms a crown on top of the stem.

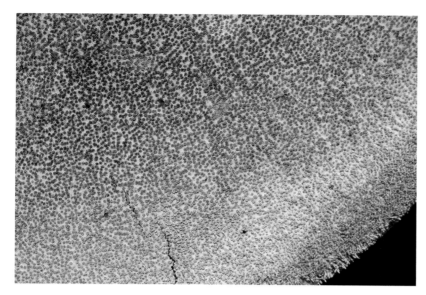

2.5 Stem cross section of the monocotyledonous Canarian Date Palm (Phoenix canariensis) from Gomera, Canary Islands. The irregular distribution of vascular bundles is characteristic for most monocotyledonous trees.

2.6 Stem cross section of a conifer. Norway Spruce (Picea abies) from the montane zone of the Alps. Seasonal climate triggers the formation of annual rings.

2.7 Stem cross section of a dicotyledonous tree. Karri (Eucalyptus diversicolor) from Southwestern Australia. The formation of annual rings is frequent in climates with mild winters and dry summers. The stem construction is principally the same as that of conifers.

Principal Construction of Roots and Shoots

Within the Pteridophyta (e.g. ferns), there is no anatomical difference between the primary meristems of roots and shoots. With the evolution of seed plants (Spermatophyta), the primary root meristem developed bipolarity, delivering cells to the rootcap and to the growing root. The primary shoot meristem retained its unipolarity, delivering cells only to the growing shoot (2.8; Huber 1961).

The xylem anatomy of all angiosperm roots remained fixed in the evolutionary stage of the Pteridophyta: the primary xylem is - as in *Psilotum nudum* - a primitive Actionostele (see p. 9). Therefore, the roots have no pith (2.9). With the evolutionary development of secondary growth, roots also underwent changes towards a more complex structure, but their center remained unchanged.

The transition zone between roots and shoots reflects not only functional but also evolutionary trends. Secondary growth in shoots of a seed plant shoots begins with the formation of vessels outside the pith. Hence, shoots have a pith (2.10). In the transition zone between roots and shoots, a complicated anatomical transition takes place.

The anatomy of roots and shoots is determined by functional differences. In contrast to roots, shoots have internodes and leaves, and their epidermis is perforated by stomata.

The anatomical differences between root and shoot xylem mainly consist of diverse vessel frequency and size, as well as different fiber diameters. All these elements are more pronounced in roots than in shoots. Whether a root or a shoot is being formed, is not genetically pre-determined: exposed roots change their structure into shoots (2.11).

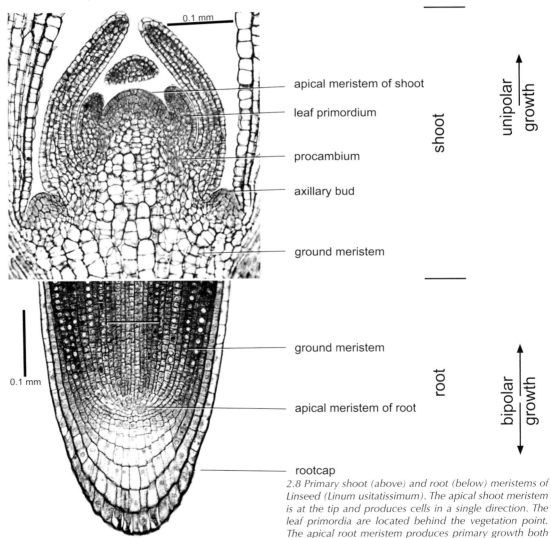

2.8 Primary shoot (above) and root (below) meristems of Linseed (*Linum usitatissimum*). The apical shoot meristem is at the tip and produces cells in a single direction. The leaf primordia are located behind the vegetation point. The apical root meristem produces primary growth both towards the root cap and the root itself (Esau 1977).

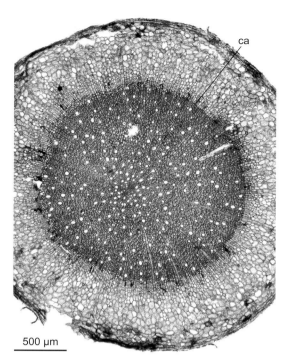

2.9 Cross-section of a Medlar root (Mespilus germanica). The missing pith is characteristic of the root.

2.10 Cross-section of a Medlar shoot (Mespilus germanica). The presence of a pith is characteristic of the shoot.

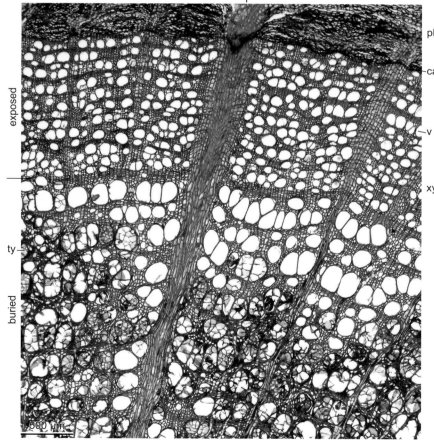

2.11 Cross-section of the exposed root xylem of a Beech (Fagus sylvatica). When a root remains in the soil, it forms a root-like structure, but as soon as it becomes exposed, its structure changes towards smaller cells.

2 THE STRUCTURE OF THE PLANT BODY

33

Principal Construction of the Xylem and Phloem
Cell Types, Cell Walls and Cell Contents

All cell types in the wood and the bark are derivates of the cambial initial cells (2.12). Secondary growth is a result of the activity of the vascular cambium (2.13). Cambial initial cells divide into xylem and phloem mother cells. They are of spindle-like form (fusiform initial cambial cells; 2.14). The result of cell differentiation and cell wall growth are longitudinal extended tracheids and fibers, cubic parenchyma cells (2.15), sclerenchyma cells (2.16), vessels (2.17, see also 3.13, p. 44) and sieve cells (2.18).

As long as cells are alive, cell walls of the xylem get differentiated in three layers: the primary wall (S1) is characterized by irregular and the secondary wall (S2) by regular orientated microfibrils (2.19, 2.20); the tertiary walls (S3) often develop warts.

Lateral communication between xylem cells is guaranteed by pits (see 3.14d, pp. 44 and 45), and between phloem cells by sieve areas (2.18). Pits in parenchymatic cells are simple (see 3.17, p. 45), in tracheids and vessels bordered (see 3.17 and 3.18, p. 45). Efficient axial communication from vessel to vessel in the xylem is possible by perforations (see 3.13, p. 45) and by sieve plates in phloem cells (2.18).

2.12 Derivates of cambial initial cells. The xylem of conifers contains tracheids and parenchyma. The xylem of deciduous trees contains tracheids, fibers, vessels and parenchyma. The phloem of conifers and deciduous trees contains sieve elements, parenchyma and sclereids (after Wagenführ 1989).

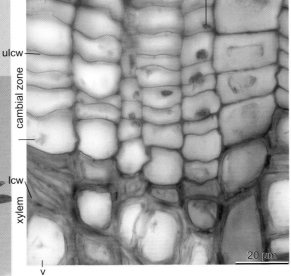

2.13 Cross section of cambial cells with nuclei and few xylem cells. Cambium initial cells cannot be distinguished from xylem or phloem mother cells. Walnut (Juglans regia).

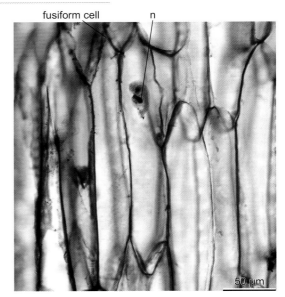

2.14 Tangential aspect of cambial cells. The fusiform cells with spindle-like shape contain nuclei. White Bryony (Bryonia dioeca).

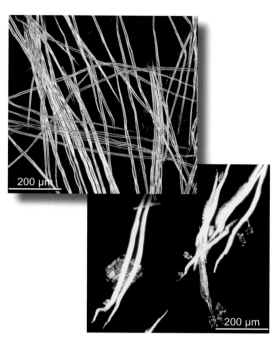

2.15 Top: macerated fibers from Scots Pine wood (Pinus sylvestris). Below: macerated fibers (long cells) and rectangular parenchyma cells. Box Wood (Buxus sempervirens, both polarized light).

2.16 Sclerenchyma cells in Walnut bark (Juglans regia). Characteristic are the very thick cell walls penetrated by openings of simple pits. The lumina are filled with tannins (red).

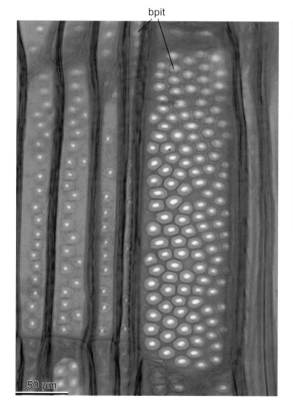

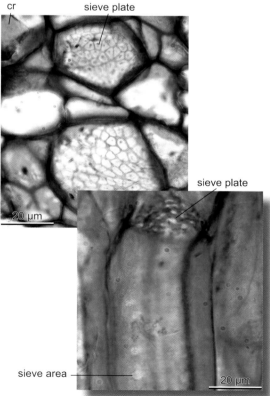

2.17 Small and wide vessels with bordered pits. Red-berried Elder (Sambucus racemosa).

2.18 Sieve cells of White Bryony (Bryonia dioeca). Top: transversal section. Below: radial section. Characteristic are the net-like sieve plates at the axial ends and sieve areas at the longitudinal sides. *continued next page*

2 The Structure of the Plant body

Liquid exchange gets blocked by the formation of sack-like structures called tyloses (2.21, 2.22). Many species develop ring- and spiral-like wall structures (2.20; see p. 44). Mechanical stress often induces reaction wood: compression wood in conifers (see p. 54) and tension wood in angiosperms (see p. 55). The living cell excretes substances: starch grains (2.23), fat, slime (2.24), resin (see p. 61) and crystals (2.25) of different forms mainly occur in parenchymatic cells. Waxes are characteristic for epithelial cells, e.g. the cuticule (2.26). Tannins are mainly produced in aging cells and are deposited in all cell types (2.27, 2.28). All these cell types are descendants of the cambium, xylem and phloem (2.29).

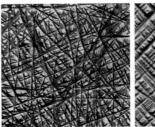 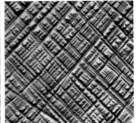

2.19 Electron microscopic photograph of microfibrils of primary (left) and secondary walls (right). Characteristic is their orientation: dispersed in primary walls (S1), parallel in secondary walls (S2) (Strasburger et al. 2002).

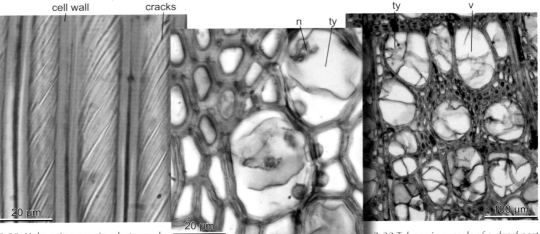

2.20 Light microscopic photograph of macrofibrils of secondary walls (S2) in the latewood of a Norway Spruce tree (Picea abies). The fibrils are spiral-like orientated.

2.21 Thin walled tyloses in vessels of Ivy (Hedera helix). Tyloses are formed by the plasmalemma of living parenchyma cell walls growing into the pith of neighbouring vessels (little blue bulbs). In the initial phase nuclei maintain cell wall growth of the tyloses in the vessel.

2.22 Tyloses in vessels of a dead part of Beech root xylem (Fagus sylvatica). Water transport is completely interrupted.

2.23 Starch grains in ray cells (big cells), axial parenchyma and ray cells of the living xylem of Bilberry (Vaccinium myrtillus). Characteristic are the round forms which reflect polarized light in cross form.

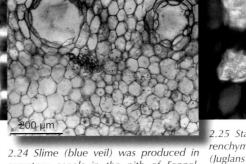

2.24 Slime (blue veil) was produced in secretory canals in the pith of Fennel-leaved Sow-Thistle, a woody Asteraceae (Sonchus leptophyllus).

2.25 Star-like crystal druses in parenchymatic cells of Walnut bark (Juglans regia).

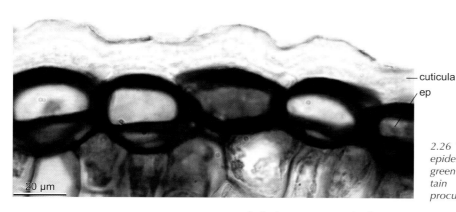

2.26 Waxes form the leaf epidermis of the evergreen dwarf shrub Mountain Azalea (*Loiseleuria procumbens*).

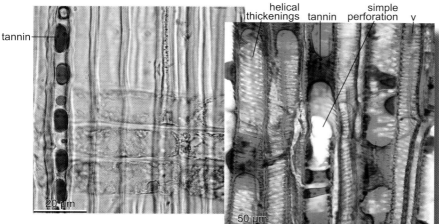

Left: 2.27 Tannins in the heartwood of a Savin Juniper (*Juniperus sabina*), visible as brown substances in axial parenchyma cells and in rays (unstained).

Right: 2.28 Tannins in vessels of the heartwood of the Fabaceae Codeso de Cumbre (*Adenocarpus villosus*). The gum-like plugs may interrupt the water flow. The vessels show simple perforations and helical thickenings.

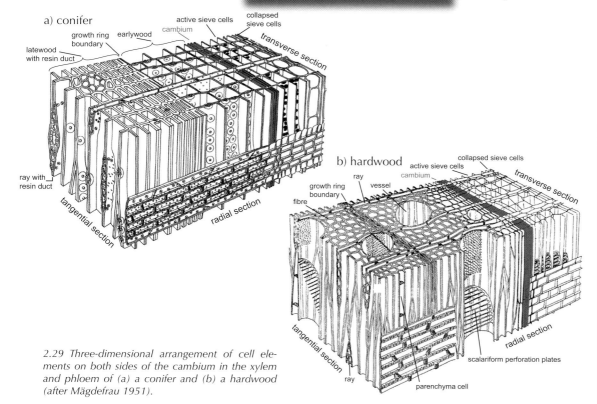

2.29 Three-dimensional arrangement of cell elements on both sides of the cambium in the xylem and phloem of (a) a conifer and (b) a hardwood (after Mägdefrau 1951).

2 THE STRUCTURE OF THE PLANT BODY

Chapter 3

Secondary Growth: Advantages and Risks

Here we explain how the cambium of different taxonomic origin optimizes anatomical structures in relation to disturbances.

Cross section of a six-year-old Stone Pine (Pinus cembra).

Primary and Secondary Growth

Growth of gymnosperms and angiosperms (monocotyledons and dicotyledons) is based on the following stages. Primary meristems, i.e. derivates from embryonic tissue, induce longitudinal growth. This is called primary growth. At the tips of all branches and roots there is a primary meristem (apical meristem) (3.1-3.3). Behind the tip, some cells remain meristematic and form the lateral meristem. This is the vascular cambium that permits secondary growth and is responsible for stem thickening (3.4). Reactivated parenchyma cells in the cortex form the periderm. A meristem of secondary origin creates the periderm. It replaces the epidermis in stems and roots, which grow bigger by secondary growth.

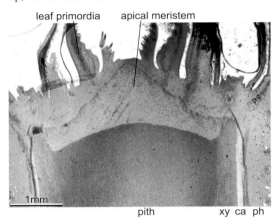

3.1 Longitudinal section of the shoot tip of an Ash twig (Fraxinus excelsior). The apical, primary meristem and the leaf primordia are shown. In the center, behind the growing region, there is the pith, and on either side of the pith the first xylem cells can be found.

Right: 3.2 Bud of an Ash (Fraxinus excelsior). Inside of the dark bud scale is the primary meristem as shown in 3.1 and 3.3.

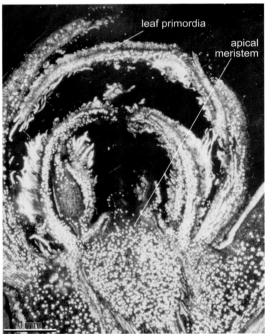

3.3 Longitudinal section of an adventitious bud on a Walnut twig (Juglans regia). Leaf primordia wrap around the primary meristem. The little white dots are probably calcium oxalate crystals (polarized light).

Right: 3.4 Simplified schematic representation of longitudinal and transversal sections of a dicotyledonous plant. The primary meristems (apical meristem and cambium) are illustrated in red. Black and green represent the derivates of the cambium, and blue the derivate of the secondary meristem (phellogen). Protoxylem, xylem, protophloem and metaphloem are not shown (after Strasburger et al. 2002).

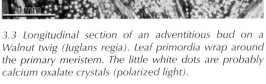

The structure of monocotyledonous and dicotyledonous plants is different.

Monocotyledons

The primary meristems form vascular bundles, which are continuous from the root tips to the stem tip. They are embedded in parenchymatous tissue. The whole tissue is surrounded by a primary, mostly sclerotized bark (3.5, 3.7).

Dicotyledons

The parenchyma cells in the pith are a product of the primary meristem. This parenchymatous tissue is surrounded by the secondary xylem, which is a product of secondary growth. The pith is surrounded by xylem and phloem, which are produced by the cambium. The whole plant body is surrounded by the periderm (3.6).

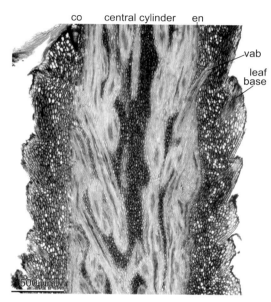

3.5 Longitudinal section of a monocotyledonous, annual Asparagus (Asparagus tenuifolium). The vascular bundles and their ends in the leaf bases are visible.

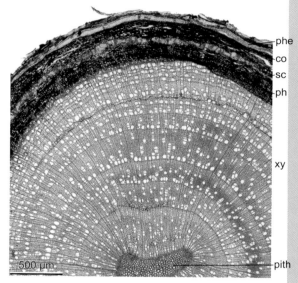

3.6 Cross-section of a perennial dicotyledonous plant twig. Silver birch (Betula pendula). Bifacial cambium growth is typical of dicotyledons. The active layer (cambium) is situated between xylem and phloem.

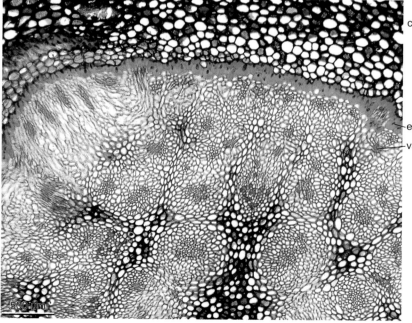

3.7 Cross-section of the monocotyledonous Scottish Mountain Asphodel (Tofieldia calyculata). The vascular bundles, which are distributed over the entire cross-section, are typical for monocotyledons. Small above: Tofieldia calyculata at the Gstettner Alm, Austrian Alps.

3 Secondary Growth: Advantages and Risks

Principle Structure of Plants with Secondary Growth

The size and distribution of all cell types are variable. This results in an immense structural variability of xylem and phloem, which depends upon individual genetic information as modified by the prevailing ecological conditions.

Transverse, tangential and radial microscopic sections illustrate species specific features. Transverse sections show the cell type distribution at different moments in time; tangential sections demonstrate variability of rays in a transverse direction; while the radial sections show cell wall construction and ray cell distribution.

All plants which are capable of secondary growth are based on the same „building plan" as demonstrated by the longitudinal and horizontal orientation of the tissue (3.8, 3.9). The transverse section illustrates the distribution and orientation of all cell types. This section shows the annual growth rings. The radial section shows details of cell wall construction and ray composition. The tangential section gives important information regarding the composition and density of the rays.

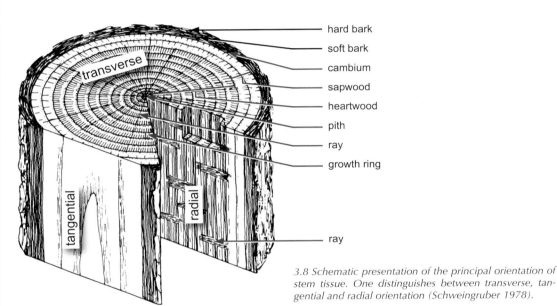

3.8 Schematic presentation of the principal orientation of stem tissue. One distinguishes between transverse, tangential and radial orientation (Schweingruber 1978).

3.9 Three-dimensional microscopic features of Acer (Core et al. 1979).

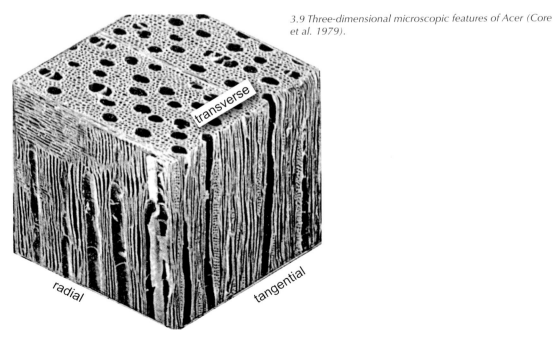

Physiological Ageing in Plants with Secondary Growth

The ecological consequence of secondary growth is the fact that plants can grow old! The ability to survive intensive stress is the basis for longevity. Since primary and secondary growth, as well as longevity, are closely related to ontogenetic and physiological processes (3.10).

All immediate derivates from the apical meristems and the cambium are physiologically young. All older products, xylem and phloem, have a different ontogenetic age; tissues with different calendar dates have different ages in relation to the seed stage but are similar with respect to the time since cell division of the meristem (3.11).

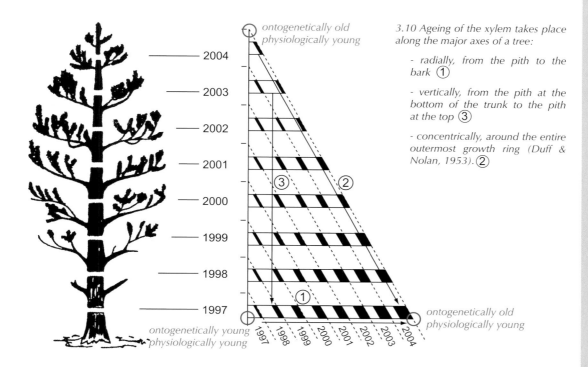

3.10 Ageing of the xylem takes place along the major axes of a tree:

- radially, from the pith to the bark ①

- vertically, from the pith at the bottom of the trunk to the pith at the top ③

- concentrically, around the entire outermost growth ring (Duff & Nolan, 1953). ②

3.11 Cross-section of a six-year-old Stone Pine stem (Pinus cembra) with all principal tissues. The pith and the cortex (below the cork layers) are derivates of the apical meristem. Both tissues are ontogenetically and physiologically young. The ring number 6, the cambium and the innermost phloem are physiologically young (one year) and ontogenetically old (6 years).

3 Secondary Growth: Advantages and Risks

The Risks of Water Transport:
Stabilized and Permeable Cell Walls

The first plants on land had already developed a system for transporting water and assimilates (Henes 1959). In order to avoid the risk of cell collapse, the walls of water-conducting cells were stabilized (3.12-3.14). Vessel cell walls of the Devonian plant *Rhynia* are ring-like and spirally thickened. Plants from the middle Devonian already possessed a net-shaped wall structure. The development of bordered pits dates back to approximately 350 million years ago. Since then, only conifers have maintained this particular system; the tracheids contain bordered pits mostly along the radial cell walls. The bordered pit torus acts as a valve. When cell pressure decreases, for example after an injury, the valve is pressed at the border of the pit opening and, thereby, seals the cell (3.15-3.18).

In a next step of evolution, the water transport system of angiosperms formed vessels with often large diameters and perforated ends (3.13). Scalariform perforations seem to be a relic of earlier evolutionary stages. Most angiosperms have simple perforations. The primitive pitting of the longitudinal walls has survived in all angiosperms to the present day.

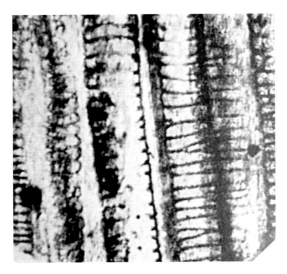

3.12 Circular cell wall thickening in an early Devonian plant (Asteroxylon mackiei; Henes 1959).

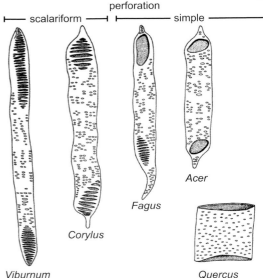

3.13 Hypothetical phylogenetic development of perforations. The perforations in Ephedra are pit-like (see 4.55, p. 81), those in Betulaceae and many others are scalariform, and in most other angiosperms they are simple (Henes 1959).

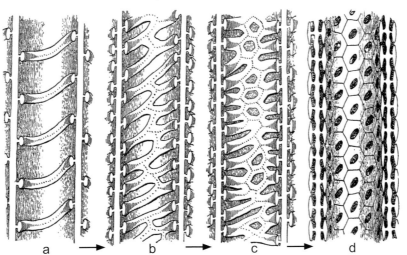

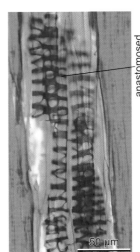

3.14 Hypothetical phylogenetic development of tracheid cell walls. Circular strips (a) anastomosed (b) and finally formed net-like cell wall structures (c) and bordered pits (d) (Henes 1959). Right: Anastomosed cell wall structures in vessels of the Crassulaceae Aeonium arboreum (Black Tree Aeonium). In this case, the lignified bands in the vessels that are surrounded by unlignified parenchyma cells.

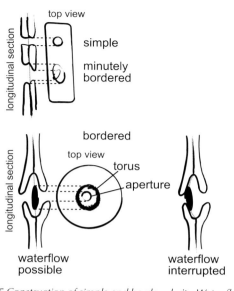

3.15 Construction of simple and bordered pits. Water flow through simple pits is always possible. Water flow through bordered pits can be interrupted by the movement of the torus to the aperture (after Strasburger et al. 2002).

Right: 3.16 Bordered pits in conifers. The tori are not lignified. See also 3.17.

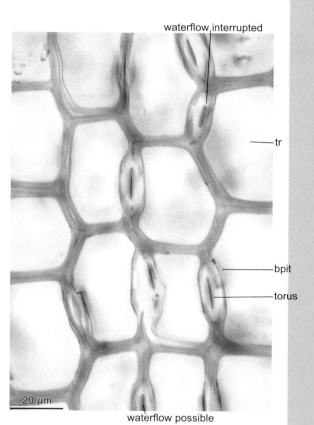

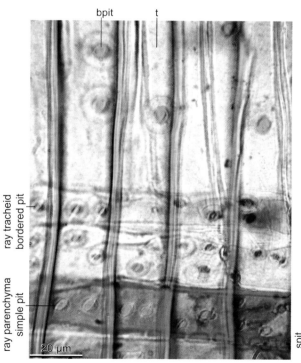

3.17 Pits in the xylem of Norway Spruce (Picea abies). Shown are the big bordered pits of longitudinal tracheids, the small bordered pits of ray tracheids and the simple pits of ray parenchyma cells (bottom). The blue dots in bordered pits represent the valves (torus); the blue dots in parenchyma cells are the thin unlignified cell walls.

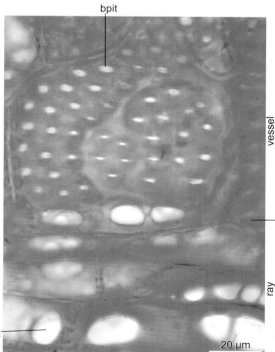

3.18 Pits in the xylem of an oak (Quercus sp.). Visible are the bordered pits on the vessel cell wall (upper half) and simple pits with big apertures in ray cells. The side walls show simple pits in the cross section.

The Risks of Stem Thickening:
Dilatation and Phellem Formation

Most cambia produce bifacially: this means that they make xylem towards the inside of the stem and phloem towards the outside. As the xylem is incompressible, the stem continually increases in circumference, with the result that the tissue inside and outside the cambium must adapt to the bigger circumference (3.19-3.21).

Anticlinal cell divisions enable the stem to enlarge its girth. At the beginning of growth, for example, a little *Pinus strobus* stem had 794 initial cells in its cambium. At the age of 60 years, the cambium had 32,000 initial cells; the number of cambial cells had, therefore, increased 90 times (Bailey 1923)! In addition, the circumference of the phloem enlarged due to the dilatation of the rays (see 3.25, 3.26, p. 48).

As the xylem thickens, the dense periderm mantle creates problems for the thin-walled phloem cells. As soon as the cell pressure (turgor) in the sieve tubes decreases, the sieve tubes are compressed. As a result, all the sieve tubes in the older phloem collapse (3.22-3.24). The pressure is released by live, inactive parenchyma cells in the outer part of the phloem which become active again and form an additional periderm while the outer periderm cracks (3.25-3.29). This new, active zone is called phellogen; centripetally, it produces a few phelloderm cells and, centrifugally, many phellem cells. The cell walls of the phellem are made of cork (suberin).

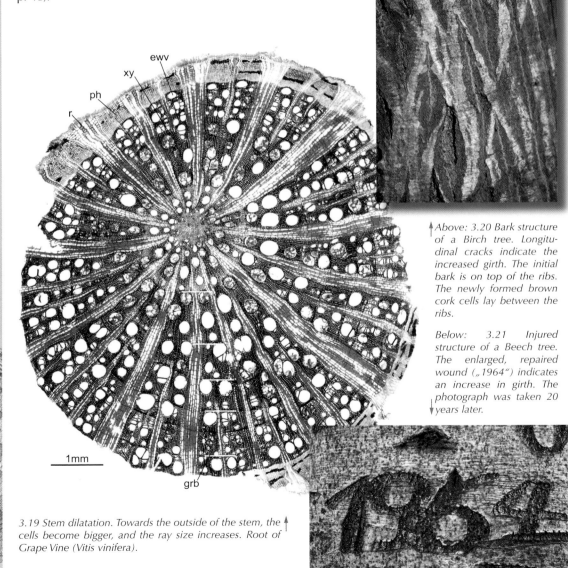

Above: 3.20 Bark structure of a Birch tree. Longitudinal cracks indicate the increased girth. The initial bark is on top of the ribs. The newly formed brown cork cells lay between the ribs.

Below: 3.21 Injured structure of a Beech tree. The enlarged, repaired wound („1964") indicates an increase in girth. The photograph was taken 20 years later.

3.19 Stem dilatation. Towards the outside of the stem, the cells become bigger, and the ray size increases. Root of Grape Vine (*Vitis vinifera*).

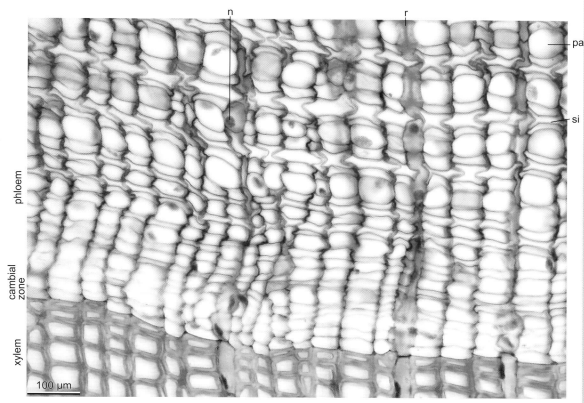

3.22-3.24 Collapse of sieve tubes and enlargement of parenchyma cells in the phloem.
3.22 The very thin-walled sieve tubes are slightly deformed. The parenchyma cells are round. Common Juniper (Juniperus communis).

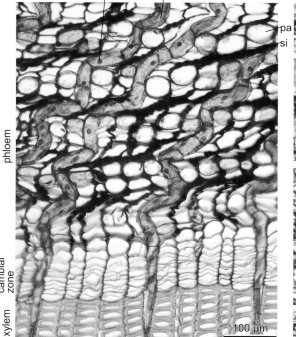

3.23 Completely compressed sieve tubes in Norway Spruce (Picea abies). The reduced diameter is indicated by the bent rays. The big, round parenchyma cells and the ray cells contain nuclei and are therefore alive.

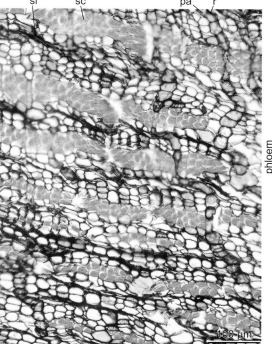

3.24 Completely compressed sieve tubes between sclerenchyma fibers in living Birch phloem (Betula pendula).

continued next page

3 SECONDARY GROWTH: ADVANTAGES AND RISKS

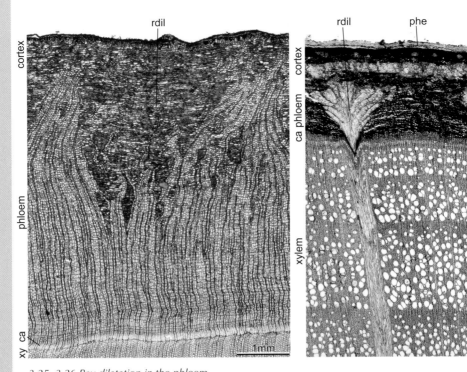

3.25, 3.26 Ray dilatation in the phloem.
3.25 Uniseriate conifer rays enlarge by forming new parenchyma cells. The beginning of dilatation occurs at different times. The newly formed parenchyma cells are alive. The former girth can still be distinguished by the phloem-wedges towards the outside. Outeniqua Yellowwood (*Podocarpus falcatus*).

3.26 Only very large rays are dilated. Immediately after the formation of the filling cells, they turn sclerotic (parenchyma cells). These parenchyma cells are able to resist the mechanical pressure of the sclerenchymatic belt below the periphery. Note the cone-like indentation. European Beech (*Fagus sylvatica*).

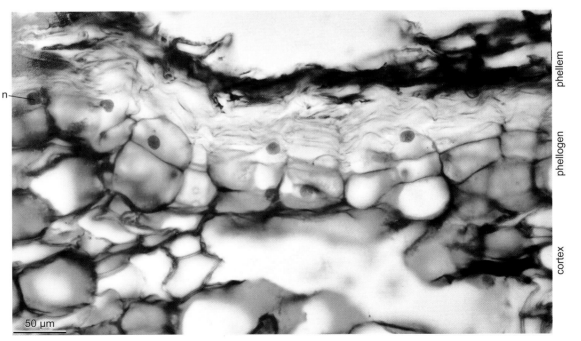

3.27-3.29 Formation of the bark zone, the periderm.
3.27 A few parenchyma cells in the cortex are capable of turning meristematic again (phellogen, cells with nuclei), thereby producing cork cells towards the outside (phellem, collapsed cells) and a few parenchyma cells towards the inside. Medlar (*Mespilus germanica*).

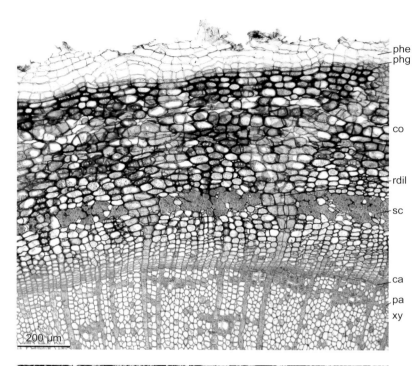

3.28 A few rows of rectangular, thin-walled cork cells cover the zone of the cortex and the phellogen. The cork cells are produced from the cork cambium (phellogen). See also the ray dilatation, thin-walled parenchyma cells break the red sclerenchymatic belt. Common Ash (Fraxinus excelsior).

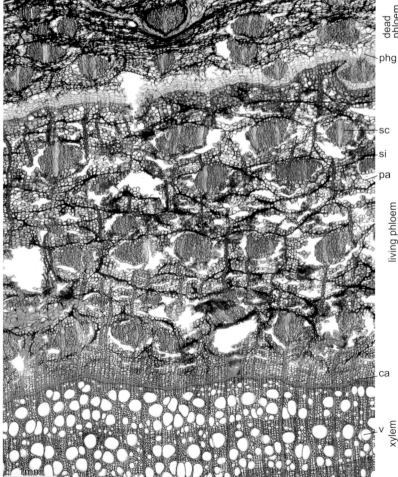

3.29 Renewed cork cambium. The increasing xylem diameter periodically triggers the formation of a new cork cambium. This example demonstrates how the live, blue/red phloem is separated from the dead, brown bark. The older cork cambia are initiated inside the phloem. Sea Buckthorn (Hippophae rhamnoides; small above: from Aeschimann et al. 2004).

3 SECONDARY GROWTH: ADVANTAGES AND RISKS

The Risks of Over-Production: Programmed Cell Death

A big plant contains many more cells than a small one. The plant must compensate for this increased cell production with a more efficient metabolism. One possible way is to increase the supply of carbohydrates is the production of more photosynthetically active cells. However, this would not be sufficient to compensate for the even faster increase in non-photosynthetic cells due to the separation of functions of roots, stems and leaves. One solution to the problem of maintaining a positive carbon balance is a differentiation between cells with a purely physical function to maintain the structure from those with a biological function of growth and assimilation. Even in the earliest ontogenetic stages, very soon after cell formation, the programmed cell death (apoptose) determines which cells may survive and which will have to die. In general, parenchyma cells survive much longer than water-conducting tracheids and vessels (3.31-3.33).

The process of programmed cell death has many time dimensions: the earlywood tracheids of conifers survive for only a few days, whereas parenchyma cells of the same wood are able to survive for decades (3.32). The live cells in the sapwood, approximately 3-5%, maintain the tree's metabolism. However, the programmed cell death also reaches the parenchyma cells. This is the moment when heartwood appears (3.30). The heartwood has no longer any metabolism, because all of its cells are dead, but its function is to maintain the tree structure.

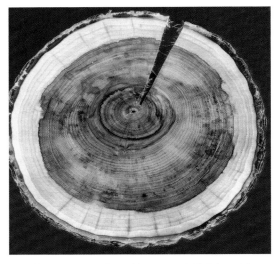

3.30 Stem cross-section with sapwood (light) and heartwood (dark brown). The heartwood is dead and it is already occupied by a wood decomposing fungus. Only the sapwood contains live cells and acts as water-conducting tissue. Black Walnut (Juglans nigra).

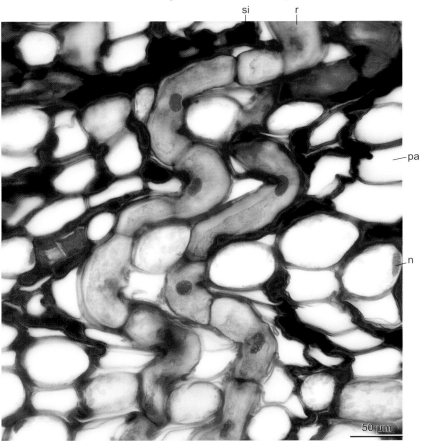

3.31 Live parenchymatous ray in the phloem. The cells with red nuclei are alive. Norway Spruce (Picea abies).

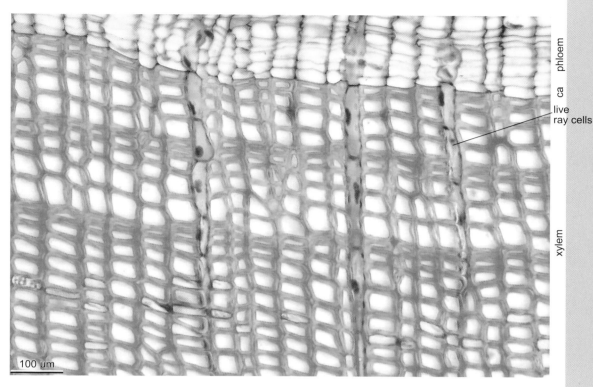

3.32 Live ray cells in the xylem. Only the parenchymatous ray cells contain nuclei. All the tracheids are dead. All the cambial cells are alive, but the nuclei disappeared during the preparation of the slide. Juniper (*Juniperus* sp.).

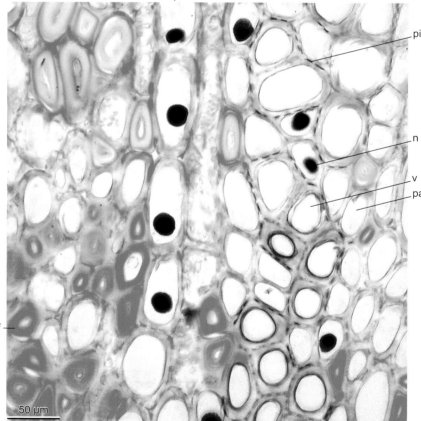

3.33 Live parenchyma cells in mistletoe xylem. The nuclei, which are present in all parenchyma cells, are extremely large. From a morphological point of view, the rays and the longitudinal parenchyma cells are hardly distinguishable. Libriform cells are thick walled and vessel cell walls are perforated by pits. White-berried Mistletoe (*Viscum album*).

3 Secondary Growth: Advantages and Risks

51

The Risks of Instability:
Eccentricity

As soon as tree stems increase in girth and height, their volume, weight and the danger of physical instability increase. The plant compensates the stress caused by weight imbalance in three ways. All seed plants are able to produce more cells on the loaded side. If the imbalance persists for any length of time, the stem becomes eccentric (3.34); if the imbalance persists only for a short period, the tree-ring pattern becomes irregular (3.36).

Enhanced stress also affects the process of cell wall formation. In general, trees produce reaction wood. Conifers respond with the formation of compression wood (3.34, 3.36), whilst many deciduous plants produce tension wood (3.35; Timell 1986).

3.34 Cross-section of a leaning Norway Spruce (Picea abies). The concentric rings in the center reflect the upright position of the tree. The eccentric phase with the compression wood formation (brown zone) indicates mechanical stress. Above small: Spruces in an avalanche track. After the event the leaning trees grew upright and stabilized themselves by forming compression wood at the lower side of the curved stem.

3.35 Microscopic section of a stem with eccentric growth. The first ring is eccentric towards the lower side, the following three towards the upper side (arrows). The last ring is incomplete. Hemicryptophytic herb, Purple Loosestrife (Lythrum salicaria) in a bog. Note the ray dilatations. Small right: from Aeschimann et al. 2004.

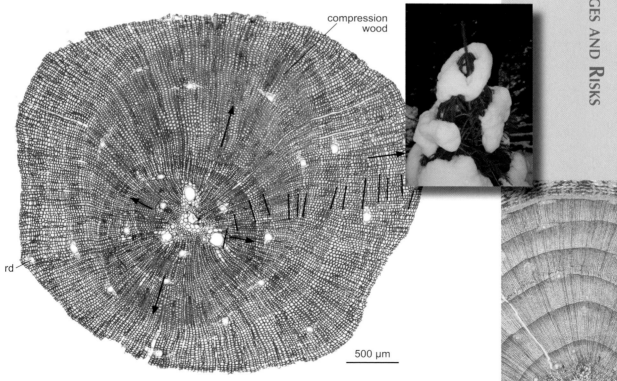

3.36 Microscopic section of a very small Mountain Pine (Pinus mugo) with eccentric rings and compression wood. The tree grew in a bog. Eccentricity and compression wood formation indicate different directions of mechanical stress (arrows). Small right: Heavy snow load on a spruce. In response to the temporal variable distribution of loads the tree builds reaction wood zones at different parts of the stem.

3 Secondary Growth: Advantages and Risks

The Risks of Instability:
Reaction Wood

Compression wood forms on the loaded side and is combined with enhanced cell production. A typical compression wood cell is circular in cross-section, has helical cavities in the cell wall structure, a high lignin content and high compression resistance. Between the compression wood cells, intercellulars are found (3.37-3.40). Compression wood is rare in roots.

3.37-3.39 Microscopic sections of compression wood in a conifer stem. Norway Spruce (Picea abies). The tree leaned over after it had been hit by an avalanche.

Small below: 3.40 Leaning Dahurian Larch trees (Larix gmelinii) on an eroded riverbank. By dating the beginning of the compression wood formation on the lower side of the trunk, it can be determined when the leaning process started. Moma River, Eastern Siberia. →

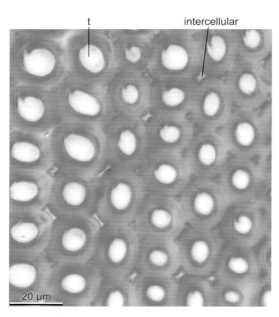

3.37 Detail of the 2001 tree ring with compression wood. The cross-section illustrates the thick round cells and their thick walls with few intercellular cavities.

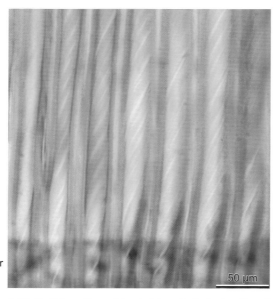

3.38 Radial section of tracheids with helical cell wall cavities.

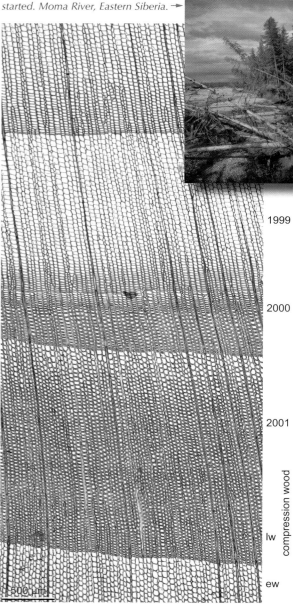

3.39 The disturbance is datable by determining the first ring with compression wood. The first affected tree ring (year 2000 after an avalanche in the winter of 1999/2000) is narrow, because the tree crown had been damaged. Despite reduced reserves, the tree was able to produce compression wood. The following two rings consist mainly of compression wood.

Tension wood is formed in approximately 50% of all tree and shrub species (Höster and Liese 1966). It also occurs in some annual shoots of herbs. Tension wood is mostly found on the tension side, i.e. on the upper side of leaning stems (3.41-3.44).

Tension wood cells are characterized by gelatinous fibers with a very high cellulose and small lignin content. Physical strength is maximized during tension. Tension wood is often present in roots.

3.41-3.43 Microscopic sections of tension wood in a deciduous tree stem of Silver Birch (Betula pendula). The tree leaned over as a result of an avalanche in the winter of 1998.

3.44 Leaning deciduous plant (Willow, Salix sp.) on a landslide. Tension wood is formed on the upper side of the trunk. The event is datable by determining, microscopically, the beginning of the tension wood formation.

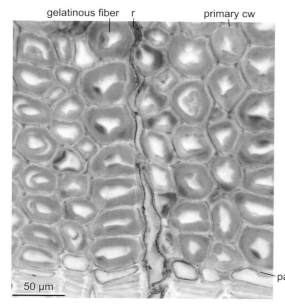

3.41 Detail of the 1999 ring. The earlywood cells are thickened with gelatinous fibers, which is typical for tension wood.

3.42 Longitudinal section of a tension wood zone. The primary cell walls are stained red, the gelatinous fibers are blue. The large fraction of primary cell walls indicates reduced lignification.

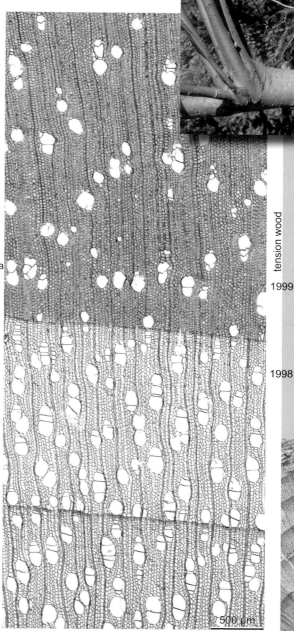

3.43 The disturbance is datable by the first ring with tension wood (blue zone). Tension wood started to form in the growing season of 1999. The avalanche took place in the previous dormant season.

The Risks of Instability:
Formation of Lignin and Thick Cell Walls

The step on land gave rise to the formation of lignin and secondary walls. This fundamental development has proved to be indispensable for all terrestrial plants, from the Devonian until the present day. Lignin is a high polymer substance, which is deposited in the interfibrillar cavities of the cellulose scaffolding of cell walls. Lignin stabilizes the cell wall.

Secondary wall formation and lignin incrustation are two processes that usually go hand in hand. Lignification is indispensable for vessels and fibers in the xylem (3.45). Its formation is optional in parenchymatous cells (3.46, 3.47). However, accelerated cell wall growth and lignification of parenchyma cells occur in the phloem, when sclereids (3.48, 3.49) are produced. The compression wood in conifers is a special form of secondary cell wall growth and lignification.

Right: 3.45 Continuous lignification of cells shortly after formation by the cambium. The unlignified (blue) cells just below the cambium - which is not visible - are the result of cell formation, their differentiation and enlargement. Cell maturation is characterized by continuous cell wall thickening and lignification (red). Common Oak (Quercus robur).

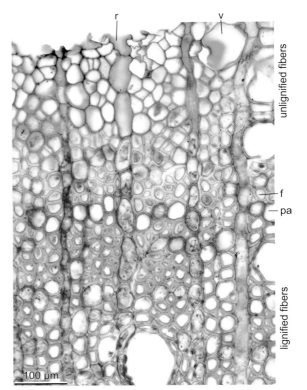

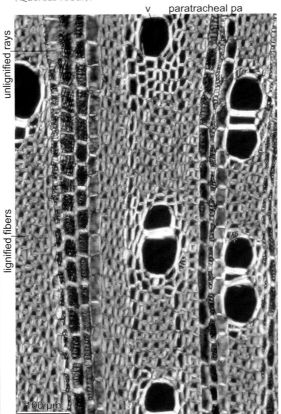

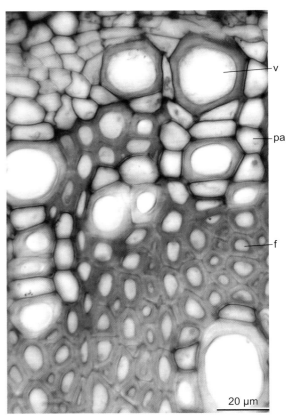

3.46 Variable cell wall thickening and lignification. Rays and cells around the vessels of Fennel-leaved Sow Thistle (Sonchus leptocephalus; polarized light).

3.47 The cell walls of parenchymatic cells are unlignified (blue). Hoary Alison (Berteroa incana), a hemicryptophytic herb.

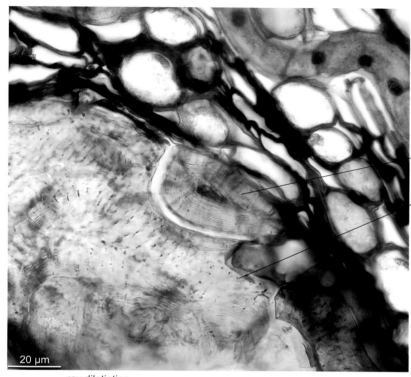

3.48 Sclereids in Norway Spruce phloem (Picea abies). The former parenchyma cells exhibit excessive expansion and cell wall growth. The periodic wall development is indicated by the fine layers, whereas the simple pits point to a parenchyma cells origin.

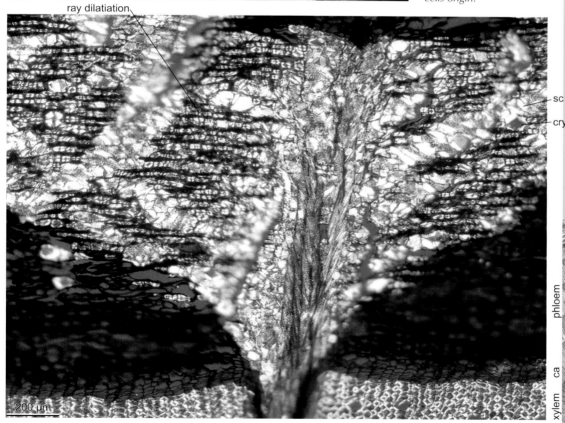

3.49 Sclereids in the dilatation of Beech bark (Fagus sylvatica). The pressure on the dense cell wall of the sclereids is much greater than the turgor in the thin-walled cells. The latter have been compressed due to sclerotic dilatation (polarized light).

3 SECONDARY GROWTH: ADVANTAGES AND RISKS

The Risk of Instability:
Internal Optimization

Every terrestrial plant is exposed to mechanical stress. Stress resistance is based on the following principles (Mattheck and Kübler 1995):

- Trees minimize external mechanical stress by reducing the length of the loaded parts (3.50, 3.51).
- Unavoidable stress is distributed evenly along the tree's surface so that it is, at least over a long time-period, uniform in every place.
- Shear stress is reduced by arranging vessels, fibers and rays along axial or tangential flow forces (3.52-3.55).
- The strength of the wood depends on how mechanical stress is distributed along the tree, when it is exposed to external loads (reaction wood).
- Weak points, such as scars, are structured in such a way that they can optimally resist the flow forces (3.56, 3.57).

3.50 This small tree lost its balance because of soil erosion. After another erosion event, apical growth continued in a vertical direction. The curved stem base demonstrates this adaptation. Norway Spruce (Picea abies), Alps, Switzerland.

3.51 Injured alpine Stone Pine (Pinus cembra, Zermatt, Switzerland). The external load on the branches was reduced by changing from horizontal to vertical growth.

3.52 Due to variable shear stress, the fibers follow axial flow forces. They actually follow the shape of the branch, „flowing" around it. Dead Mountain Pine (Pinus mugo).

3.53 Two main Sweet Chestnut tree shoots (Castanea sativa) produced forks. At the base of the fork, the fibers are closely interwoven. But soon afterwards, they changed to a vertical direction.

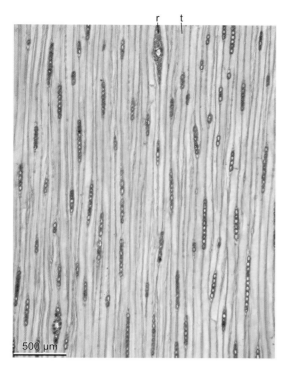
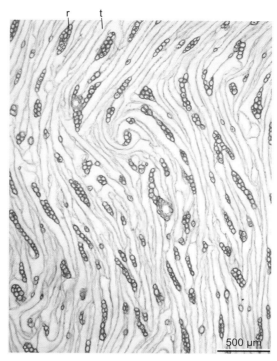

3.54 Small, slender, parallel rays are characteristic of the xylem of straight, upright-growing stems. Norway Spruce (Picea abies).

3.55 Irregularly arranged large and short rays indicate irregular flow forces. Note the biseriate rays. Norway Spruce (Picea abies).

3.56 Overgrown wounds (scars) on Cottonwood trees (Populus sp.) in the riparean zone of the St. Laurence River in Quebec, Canada. The longitudinal shape of all scars indicates the main flow of forces.

3.57 De-barked Sweet Chestnut stem (Castanea sativa) with a branch and a scar. The longitudinal shape of the scar, as well as the fiber direction around it and the branch, result from the principal flow of forces. Ticino, Switzerland.

The Risk of Decomposition:
Natural Boundaries and Protection Systems

Longevity increases the risk of injuries and attack by decomposers. The entire plant body is attractive to decomposers, mainly fungi. Young, live cells contain many easily decomposable substances, such as sugar, starch and proteins, whilst old, dead cell walls are a nutritional basis for many fungi. In stems with coloured heartwood, the demarcation lines of chemically protected zones with phenolic substances are easily visible. Dead stems in the forest, and old construction timber, clearly demonstrate that heartwood is more resistant towards fungi and insects than sapwood (3.58, 3.59).

Resin protects plants from decay. Conifers, in particular, developed a special duct system, which produces resin (3.60-3.62). Many dicotyledons produce toxic substances, e.g. phenolic substances to prevent fungal growth (3.63, 3.64).

3.58 Scots Pine stem (Pinus sylvestris) with decayed sapwood and well-preserved heartwood.

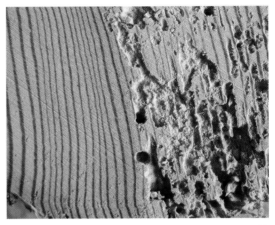

3.59 Larch beam (Larix dedcidua) removed from a building. The sapwood was attacked by fungi and insects and decayed. Three hundred years after the tree was felled, the heartwood is still well preserved.

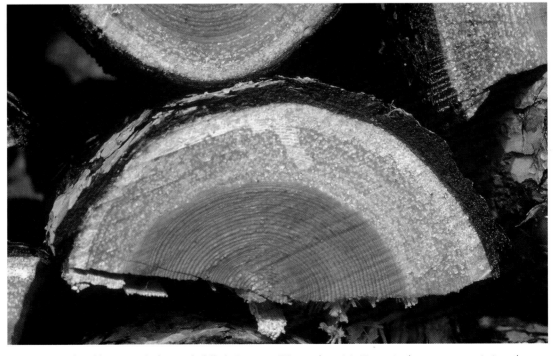

3.60 Sapwood and heartwood of recently felled pine trees (Pinus sylvestris). Live resin ducts excrete resin into the sapwood and reduce the immediate danger of fungal decay.

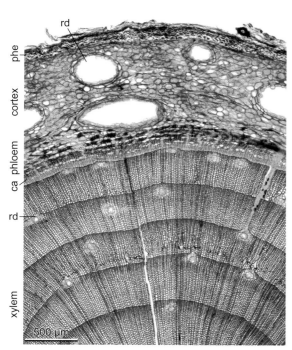

3.61 Cross-section of a Stone Pine twig (Pinus cembra). The big resin ducts in the cortex and the xylem are particularly noticeable.

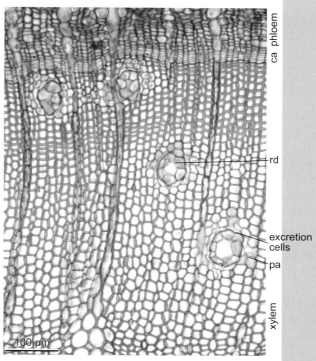

3.62 Cross-section of a Stone Pine twig (Pinus cembra) near the cambium and the xylem. The surrounding resin duct cells are alive and contain nuclei. The excretion cells produce resin and excrete it into the ducts.

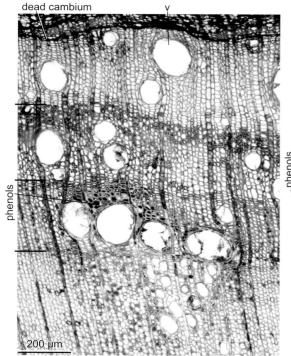

3.63 Phenolic substances in fibers of the barrier zone in a Sweet Chestnut stem (Castanea sativa). The substances inactivate the vessels and are toxic to fungi.

3.64 Phenolic substances in fibers and vessels of the barrier zone in a Red Bearberry stem (Arctostaphylos uva-ursi).

The Risk of Decomposition:
Defence Barriers around Wounds

Plants developed a compartmentalization system (CODIT, Shigo; 1989) in order to prevent infection by decomposing fungi after injuries. This protective system functions on the basis of chemical barriers. After an injury, the xylem produces phenolic substances which are a protection against wood decomposing organisms. The principle of compartmentalization is very old in evolution. Conifers, dicotyledonous angiosperms and even some monocotyledonous plants, from herbaceous plants to big trees, show the same reaction to injuries.

Since injuries are the predominant entering places for parasitic and saprophytic organisms, plants developed mechanisms to seal the wounds as soon as possible after the natural „skin" (epidermis and periderm) had been broken. Strong chemical barriers form along the rays and growth zones (3.65-3.67). Weak barriers exist in the axial direction. Injuries activate growth locally. As a result of this, a scar is formed (3.68). On the inside, the overgrown zone is resistant against new invasion by fungi. The newly produced tissue is able to isolate affected zones.

3.65 Two compartmentalized wounds on the stem of a big tree. Immediately after the stem had been wounded, live cells excreted phenolic substances below the injury, which acted like a fungicide. In this zone (dark area), fungal growth was then impossible, or at least retarded. Later on, the wound was overgrown by a callous margin (light area beside the wound). The barrier between the newly formed and the old wood is very strong and hardly ever breaks (Chosenia arbutifolia, a big willow in Siberia).

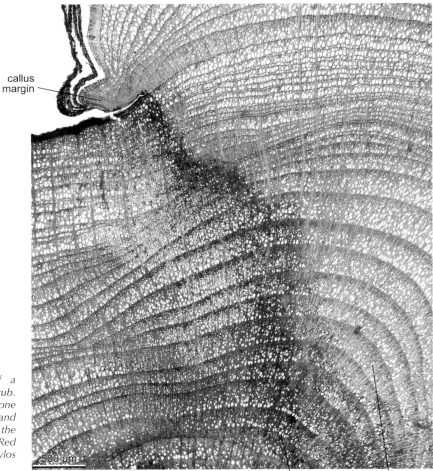

3.66 Cross-section of a wound on a dwarf shrub. Note the dark barrier zone in the radial direction and the unaffected wood in the overgrowing margin. Red Bearberry (Arctostaphylos uva-ursi).

callus margin

dead xylem — barrier zone — living xylem

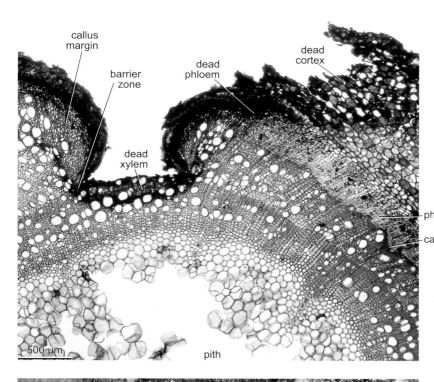

3.67 Microscopic cross-section of a wound on a biannual, herbaceous plant stem. Characteristic is the dark, compartmentalized, injured zone and the overgrowing callous margin. Horse Mint (Mentha longifolia).

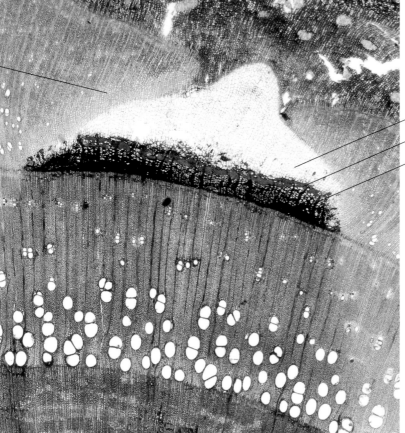

3.68 Hidden wound in an Ash stem (Fraxinus excelsior). The stem's cambium was injured without damage to the bark. The cambium and the phloem died and the xylem reacted with the formation of dark, phenolic substances. Live cells beyond the dead zone were transformed to a cork cambium and produced cork cells. During the following differentiation process, the cambium produced different tissue: xylem on the side of the triangle and phloem on its tip.

3 Secondary Growth: Advantages and Risks

The Risk of Shedding Plant Parts:
Abscission

Long-lived plants shed tissue and organs. This process is called abscission (Addicot 1981). Falling bud scales (3.70), leaves (3.69, 3.75), needles (3.70) and fruit (3.71) are the most common examples of shed plant parts. All these organs are closely attached to the stem as long as they fulfil their respective function. Vascular bundles guarantee the conduction of liquids; sclerenchymatous cells ensure stability, and parenchyma cells store mainly carbohydrates. Before programmed cell death leads to the shedding of the organ, the cork-producing phellogen is activated and builds a barrier between the stem and the respective organ (3.74). As soon as the barrier is complete, any slight mechanical stress induces the dropping of the organ. By that time, the connecting zone between the organ and the rest of the plant is completely sealed. The point from which the organ is lost is called a leaf-scale scar or bud-scale scar.

The same mechanism exists on the outside of stems and roots. Most trees, but also shrubs, dwarf shrubs and herbaceous plants, drop older compartments of the bark (3.72, 3.73). The bark pattern often varies between species, or even different clones (Vaucher 1990). Compartments break and produce a new, rough bark (phellem).

A special case is the abscission of twigs (Roloff 2001). Cottonwood (*Populus* sp.) and oaks (*Quercus* sp.) drop short shoots (Höster et al. 1968) (3.76–3.78). The abscission zone is characterized by the formation of thin-walled parenchymatous tissue (3.79) probably with calcium oxalate crystals (3.80). The anatomy of the breaking zone is very different from those in the twig (3.81, 3.82).

3.69 Leaf scars on a tree fern (Cyathea capensis). The arrangement of the sealed vascular bundles inside the leaf scar is characteristic of ferns. Garden, Madeira.

3.70 Scars of needles (short shoots) and buds on a twig of Canary Pine (Pinus canariensis). The needle scars are spirally arranged along the twig. The bud-scale scars show a small zone with horizontal structures and indicate the boundary between annual longitudinal growth increments. Forest, Tenerife.

3.71 Fruit and leaf scars on a tropical fruit tree (Papaya, Carica papaya). The small, rectangular fruit scar is just above the big leaf scar. Garden, Tenerife.

3.72 Bark-shedding on a young Dahurian Larch stem (Larix gmelinii). Boreal forest, Jakutsk, Siberia.

3.73 Bark-shedding on an old Oriental Plane (Platanus orientalis). The bark's surface is irregular, and the plates do not represent annual growth zones. City of Zürich, Switzerland.

continued next page

3 SECONDARY GROWTH: ADVANTAGES AND RISKS

The Risk of Shedding Plant Parts:
Abscission of leafs and twigs

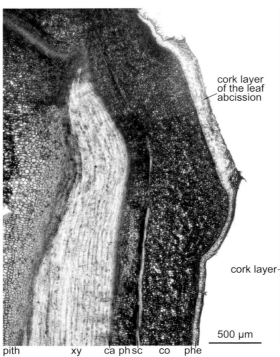
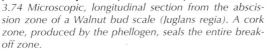
pith　　　xy　　ca ph sc　co　phe

3.74 Microscopic, longitudinal section from the abscission zone of a Walnut bud scale (Juglans regia). A cork zone, produced by the phellogen, seals the entire break-off zone.

3.75 Distinct break-off zones of leaves. The little brown zone at the base of the leaves corresponds with the cork layer as it is shown in 3.74. Sycamore (Acer pseudoplatanus).

3.76 View of a morphologically determined fracture on a short oak shoot.

3.77 Overgrown fractures (abcission zone) of short oak shoots.

Right: 3.78 Oak twigs from which the bark has been peeled off. There is a morphologically determined fracture at the base of the thinner twig (short shoot). →

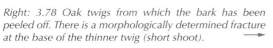

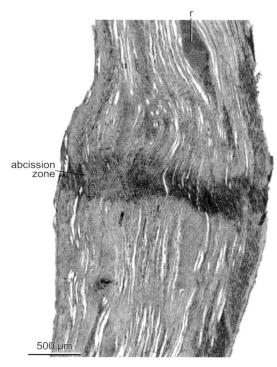

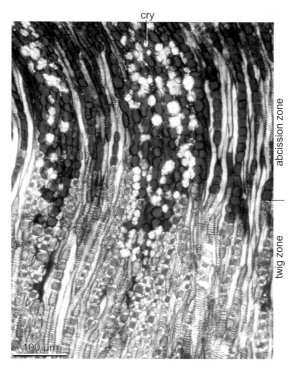

3.79 Microscopic, longitudinal section across a twig with an abscission zone. The hardly lignified tissue represents a morphologically determined fracture.

3.80 Microscopic, longitudinal section across the transition zone from the basal part and the fractured zone of the twig. The fractured zone contains many crystals (polarized light).

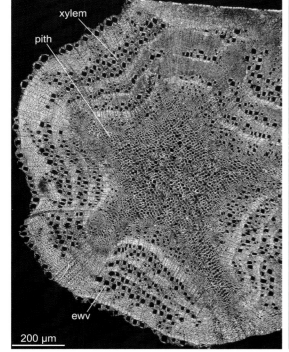

3.81 Cross-section of a short shoot. The anatomical structure is characteristic of oak: a star-shaped pith and ring-porous xylem structure (polarized light).

3.82 Cross-section of the abcission zone. The anatomical structure is different from the normal oak structure: the pith is not clearly separated from the xylem, which is diffuse-porous; the ring boundaries are indistinct, and the rays contain calcium oxalate crystals (polarized light).

Chapter 4

Modification of the Stem Structure

The anatomy of stem and bark of ferns, conifers, monocotyledons and dicotyledons is based on few growth principles. This chapter illustrates the construction and distribution of liquid conducting systems and protection elements.

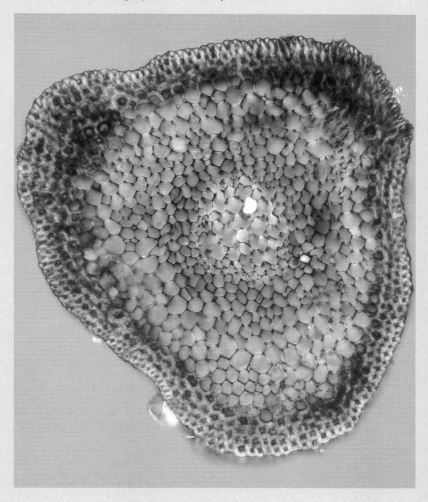

Protostele of a gametophyte stem of the Hair Cap Moss (Polytrichum commune).

The Primary Stage of Growth:
The Construction of Vascular Bundles

The first indication that a tissue conducts liquids is the presence of vascular bundles. They consist of water conducting xylem and assimilate transporting phloem (4.1). Primitive vascular bundles in mosses are composed of elongated parenchyma cells or fibers (4.2, 4.3). The cell walls of water conducting hydroids are lignified and the cell walls of assimilate transporting leptoids are not lignified (4.2). In contrast, the xylem of vascular bundles in ferns (4.4) and seed plants (conifers 4.6, angiosperms 4.5, 4.7-4.13) consist of vessels, fibers and parenchyma, and the phloem contains sieve tubes, accompanying cells and parenchyma. Generally, the bundles appear together with groups of sclereids (4.9). The primary vessels of the protoxylem demonstrate annular and those in the metaxylem screw-shaped thickenings (see 3.14, p. 44).

The arrangement of xylem and phloem is very variable. In principle, one distinguishes between vascular bundles without a cambium (= closed vascular bundles; 4.5-4.12) and those with a cambium (= open vascular bundles; 4.13). The bundles are radial where the parts are separated laterally (4.2, 4.3), for example in roots; they are concentric where the xylem surrounds the phloem or vice versa (4.4), for example in ferns; or they are collateral when xylem and phloem are located centripetally and centrifugally (4.9-4.12).

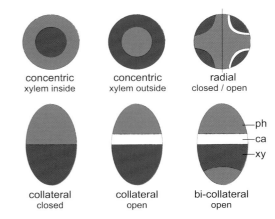

4.1 Vascular bundle types (after Strasburger et al. 2002).

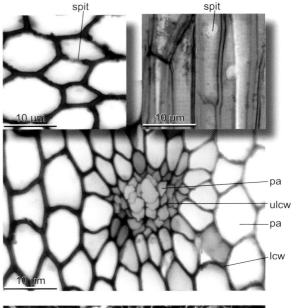

Right: 4.2 Primitive vascular bundle in the stem of the Foxtail Feather-moss (Thamnobryum alopecurum, Bryales). The thick walled water conducting hydroids surround the few assimilate conducting thin walled leptoids. The detail shows a hydroid with simple pits in transversal (small left) and longitudinal (small right) sections.

4.3 Primitive type of a closed, concentric vascular bundle. The leptoids in blue (primitive phloem) surround the hydroids in red (primitive xylem) in a moss gametophyte. Hair Cap Moss (Polytrichum commune).

4.4 Radial vascular bundle. The phloem is located outside the xylem. Whisk Fern (Psilotum nudum).

4.5 The xylem surrounds the phloem. Monocotyledonae, Mountain Scottish Asphodel (Tofieldia calyculata).

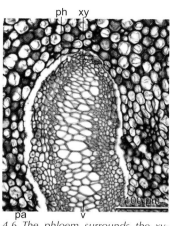

4.6 The phloem surrounds the xylem. The bundle is embedded in parenchymatic tissue. Fern, Green Spleenwort (Asplenum viride).

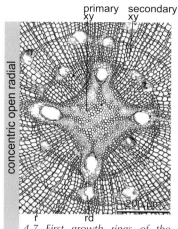

4.7 First growth rings of the root xylem. Conifer, Black Pine (Pinus nigra).

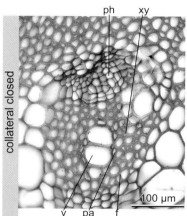

4.8 The xylem is located centripetally and the well bordered phloem centrifugally. Chenopodiaceae, Saltbush (Atriplex sp.).

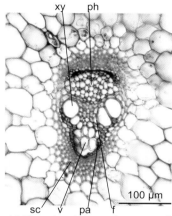

4.9 The xylem is located centripetally, the well-bordered phloem centrifugally. The bundle is surrounded by sclerotic fibers. Monocotyledonae, Pendulous Sedge (Carex pendula).

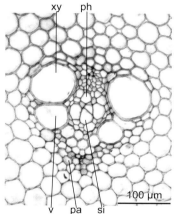

4.10 The xylem surrounds the phloem like a horseshoe. Monocotyledonae, Asparagus (Asparagus tenuifolium).

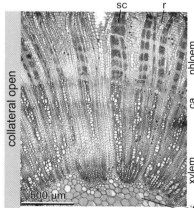

4.11 The bundles enlarge and form a closed circle. The xylem is characterized by annual and the phloem by rhythmic, non-annual growth layers. Hemicryptophytic herb, Least Mallow (Malva parviflora).

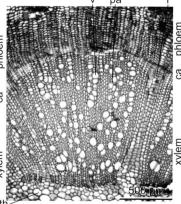

4.12 Growing mechanisms are the same as in 4.11. Hemicryptophytic herb, Whorled Sage (Salvia verticillata).

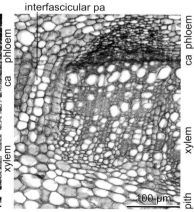

4.13 Rhizome, same growing mechanisms as in 4.11. Hemicryptophytic herb, Meadowsweet (Filipendula ulmaria).

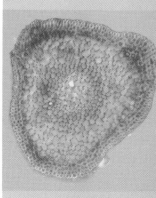

The Primary Stage of Growth: The Arrangement of Vascular Bundles in Mosses, Lycopods and Ferns

The stele (arrangement of vascular bundles in shoots and roots) can be divided into two types: the protostele (4.14-4.19) and the siphonostele (4.20-4.23). The protostele forms a solid central cylinder of xylem, consisting of one vascular bundle. The vascular bundles in a siphonostele surround the pith (4.20-4.23). Phylogenetically, a protostele is considered, and it is thought that the siphonostele developed from the protostele. In contemporary flora, protostelae are mainly found in ferns (4.17) and lycopods (4.16), siphonostelae in the shoots of seed plants. The lateral, vascular bundles in roots are relicts (4.18, 4.19).

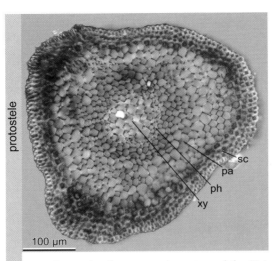

4.14 Protostele of a gametophyte stem of the Hair Cap Moss (Polytrichum commune). The central cylinder is surrounded by thin-walled parenchymatous tissue and by a belt of sclerotic fibers.

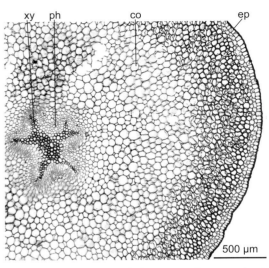

4.15 Protostele of the Whisk Fern (Psilotum nudum), similar to a lycopod. The star-shaped central cylinder (actinostele) is surrounded by very large parenchymatous tissue and by a belt of cells with lignified walls. For details see p. 9.

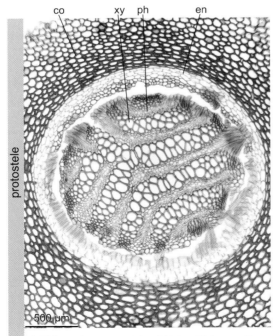

4.16 Lycophyte protostele of Alpine Clubmoss (Lycopodium alpinum). The central cylinder consists of layers of xylem and phloem (plectostele). The cylinder is surrounded by a belt of parenchyma cells and by a layer of endodermis. On the outside of the cylinder, there is the cortex.

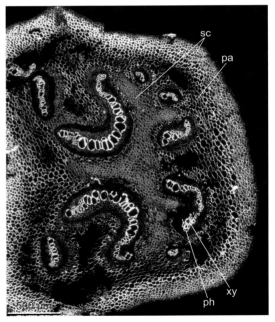

4.17 Fern polystele, Bracken Fern (Pteridium aquilinum). Sclerotic fibers form the central column of the star-shaped center. This is surrounded by several concentric vascular bundles, with xylem in their centres.

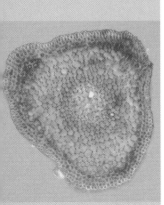

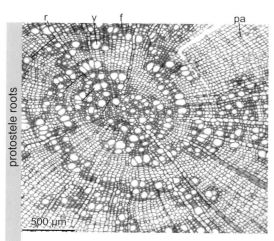

4.18 Xylem of the protostele in the root of an Angiospermae. Characteristic is the missing pith. Garland Flower (Daphne striata).

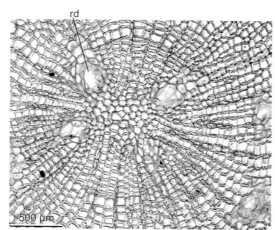

4.19 Xylem of the protostele in the root of a conifer with star-like arrangement of primary vascular bundles and resin ducts. Norway Spruce (Picea abies).

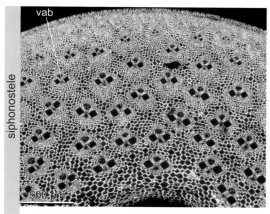

4.20 Bamboo ataktostele (Monocotyledonae, Phyllostachys edulis). The vascular bundles are arranged in spirals, following mathematical rules (Fibonacci's law).

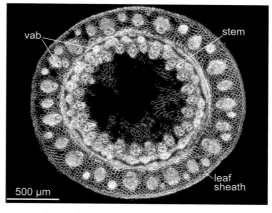

4.21 Siphonostele in a stem of a monocotelydonous plant with the vascular bundles arranged around a pith. The stem is surrounded by a compact leaf sheath. Barley (Hordeum vulgare).

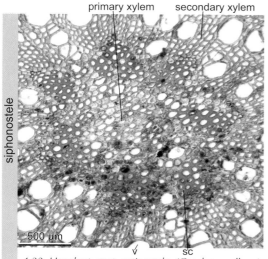

4.22 Hazelnut root actinostele (Corylus avellana). Polyarch arrangement of the primary xylem.

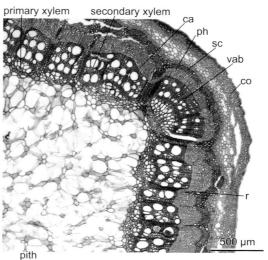

4.23 Siphonostele in a dicotyledonous plant stem. Characteristic is the arrangement of vascular bundles in circles, separated by large rays. Blackberry (Rubus fruticosus).

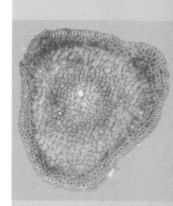

4 MODIFICATION OF THE STEM STRUCTURE

The Primary Stage of Growth: The Arrangement of Vascular Bundles in Conifer and Dicotyledonous Plant Shoots

Open, collateral vascular bundles are arranged in a siphonostele. Shortly after the formation of the bundles, meristematic tissue bridges the bundles and forms a closed cambial cylinder. There are primary rays between the bundles. Strasburger *et al.* (2002) described the shape of protoxylem and metaxylem, the arrangement of the vascular bundles, and the space between them (4.24-4.32). Pith shape is in most cases independent of the vascular bundle arrangement.

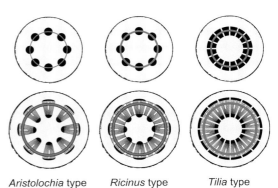

Aristolochia type   *Ricinus* type   *Tilia* type

4.24 Types of secondary growth in Eucotelydones (after Strasburger et al. 2002).

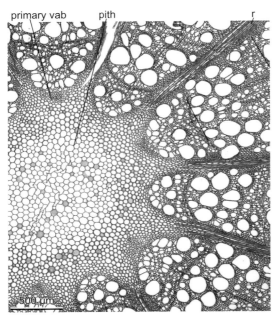

4.25 Aristolochia type. The vascular bundles are distinctly separated by large rays. Traveller's Joy (Clematis vitalba), vine.

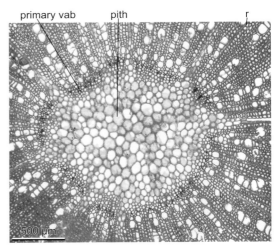

4.26 Ricinus type. The primary vascular bundles are distinctly separated. As soon as the cambium produces new tissues, the original form of the the vessels disappears. Hornbeam (Carpinus betulus), tree.

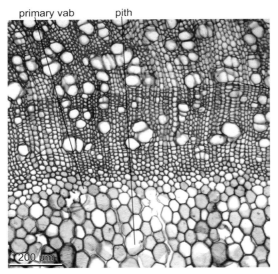

4.27 Tilia type. The primary vascular bundles are hardly separated. They form a closed xylem belt around the pith. Horse Mint (Mentha longifolia), hemicryptophytic herb.

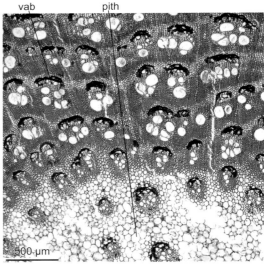

4.28 Monocotyledon-like vessel arrangement in the stem center. Before forming a complete stem, Bougainvillea (Bougainvillea spectabilis), a type with a successive cambium, forms single vascular bundles.

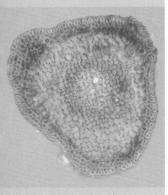

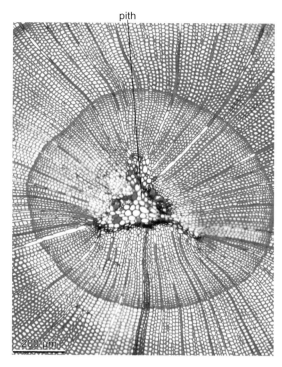

4.29 Conifer with a triangular pith. The triangular pith is characteristic of Junipers (Juniper sp.).

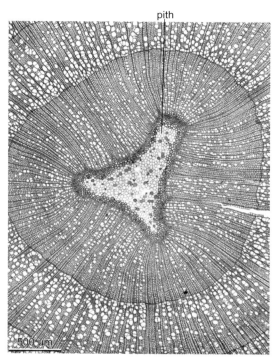

4.30 Angiosperm with a triangular pith. The triangular pith is characteristic of Alnus. The separation of the vascular bundles around the pith is indistinct. Tilia type. Grey Alder (Alnus incana), tree.

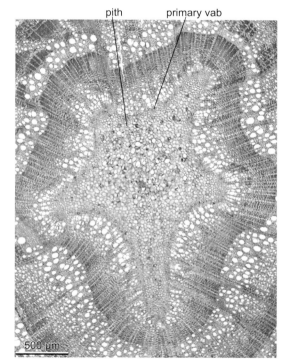

4.31 Angiosperm with a star-shaped pith, which is characteristic of Quercus trees. The separation of vascular bundles around the pith is fairly indistinct. Ricinus type. Common Oak (Quercus robur), tree.

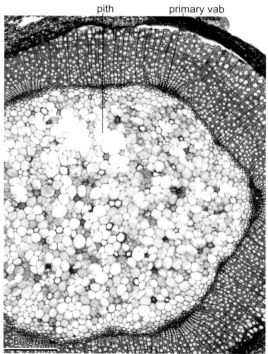

4.32 Angiosperm with a big round pith. Vascular bundles are only separated in primary stages. Wayfaring Tree (Viburnum lantana), shrub.

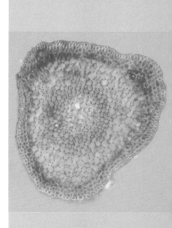

The Secondary Stage of Growth: Conifer Xylem

There is little variability in conifer xylem. The absence or presence of a coloured heartwood is distinctive for many species (4.33). All conifer species lack vessels and have exclusively uniseriate rays. Only the presence or absence of resin ducts is generally variable (4.34, 4.35).

The rays are characterized by pits of different forms and sizes in the parenchyma cells (4.36-4.38), and may or may not contain tracheids. Few species have spiral thickenings in tracheids (4.39).

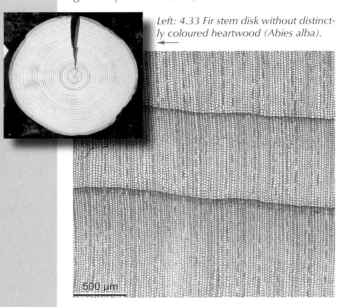

Left: 4.33 Fir stem disk without distinctly coloured heartwood (Abies alba).

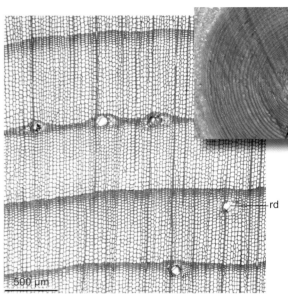

4.34 Conifer without resin ducts. Common Yew (Taxus baccata).

4.35 Conifer with resin ducts. Cross-section, Scots Pine (Pinus sylvestris), with distinctly coloured heartwood.

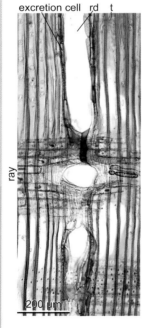

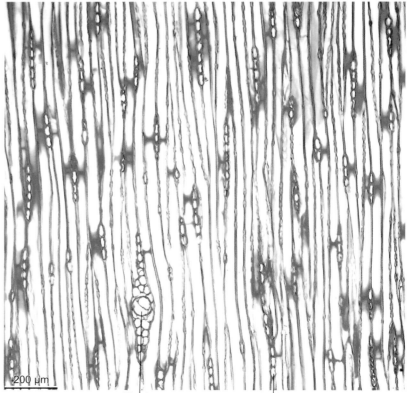

4.36 Conifers with resin ducts. Left: Radial section, Norway Spruce (Picea abies). Right: Tangential section, Scots Pine (Pinus sylvestris).

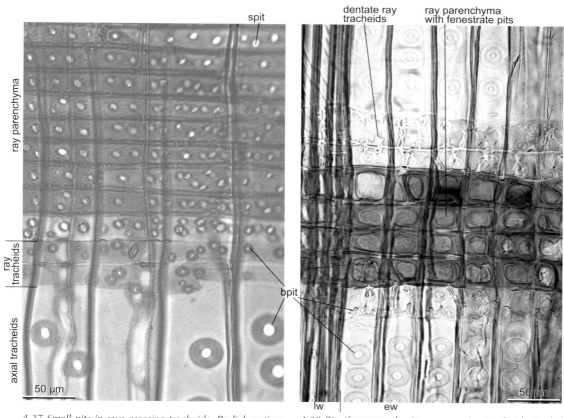

4.37 Small pits in rays crossing tracheids. Radial section, Norway Spruce (Picea abies).

4.38 Big (fenestrated) pit rays, crossing tracheids. Radial section, Scots Pine (Pinus sylvestris).

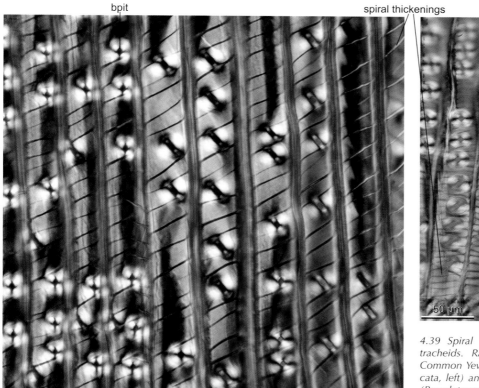

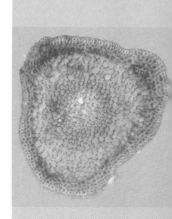

4.39 Spiral thickening in tracheids. Radial section, Common Yew (Taxus baccata, left) and Douglas Fir (Pseudotsuga menziesii, right). Polarized light.

The Secondary Stage of Growth: The Xylem of Dicotyledonous Angiosperms

The different heartwood colouring (4.40-4.42), structural features and odours allow a limited macroscopic identification of wood species. The anatomical variability is enormous, mainly due to vessels with different perforations (4.53-4.55), variable vessel size (4.44-4.46) and ray width (4.47-4.49), the presence of axial parenchyma (4.50-4.52), the variable size and form of punctuations in the ray-vessel cross-fields (4.56, 4.57), and the presence or absence of crystals.

4.40 Stem disk of a broadleaf tree without distinctly coloured heartwood. Hakea sp. Proteaceae.

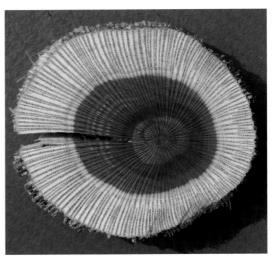

4.41 Stem disk of a broadleaf tree with distinctly coloured heartwood. Chinaberry (Melia azedarach).

4.42 Stem disk of a subfossil broadleaf tree with very distinct heartwood. Phenols got oxidized in water and turned black. This is typical for oak wood. Oak from a Neolithic lake-dwelling.

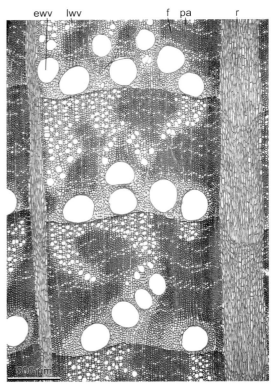

4.43 Ring-porous wood. Large rays and a patchy distribution of libriform fibers are typical for oak. Common Oak (Quercus robur).

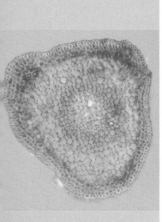

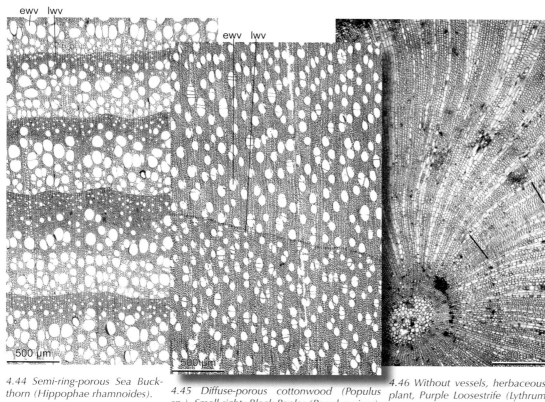

4.44 Semi-ring-porous Sea Buckthorn (Hippophae rhamnoides).

4.45 Diffuse-porous cottonwood (Populus sp.). Small right: Black Poplar (Populus nigra).

4.46 Without vessels, herbaceous plant, Purple Loosestrife (Lythrum salicaria).

4.47 Uniseriate rays, Sweet Chestnut (Castanea sativa).

4.48 Two to three rows of ray cells, Black Locust-tree (Robinia pseudoacacia).

4.49 One to numerous rows of ray cells, Common Beech (Fagus sylvatica).

continued next page

4 MODIFICATION OF THE STEM STRUCTURE

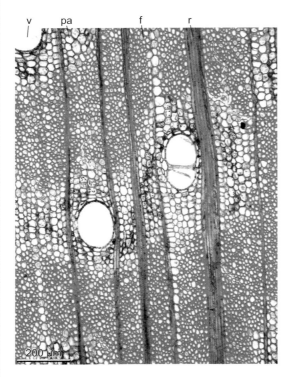

4.50 Paratracheal parenchyma: the parenchyma cells are close to the vessels. Flat-top Acacia (Acacia abyssinica).

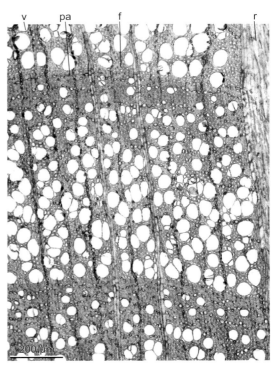

4.51 Apotracheal parenchyma, which means: the parenchyma cells are not in contact with the vessels. Single (blue) cells are widely dispersed over the cross-section. Common Beech (Fagus sylvatica).

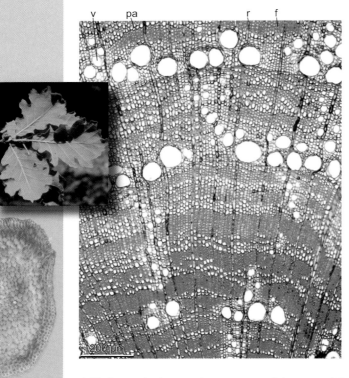

4.52 Apotracheal parenchyma, arranged in tangential bands. Downy Oak seedling (Quercus pubescens).

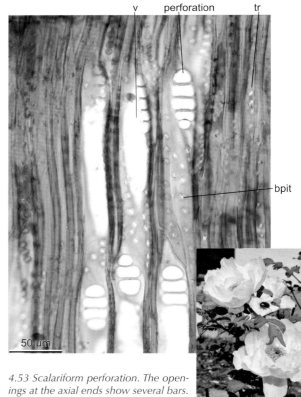

4.53 Scalariform perforation. The openings at the axial ends show several bars. Mountain Peony (Paeonia suffruticosa).
Small: Flowering Paeonia in a garden in Jena, Germany.

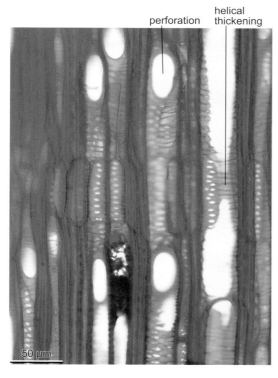

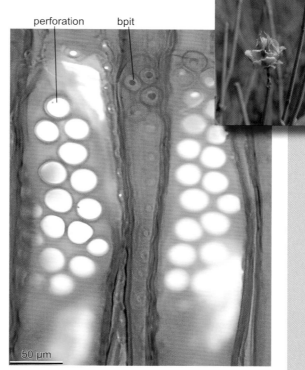

4.54 Simple perforation. The vessels have two openings at the axial ends. Pyrenean Honeysuckle (*Lonicera pyrenaica*).

4.55 Pit-like (ephedroid) perforation. The vessels show groups of pit-like openings on the radial side. Joint Pine (*Ephedra helvetica*). Small: female flower of *Ephedra alata*.

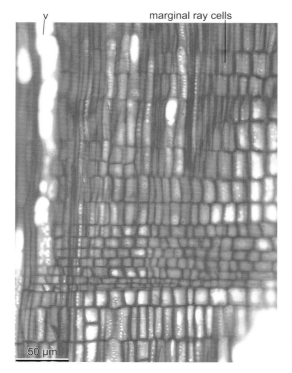

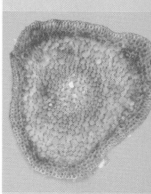

4.56 Heterogeneous rays. The axially bordering cells are in an upright position. Ray-vessel cross-fields with big pits. Red-berried Elder (*Sambucus racemosa*).

4.57 Homogeneous rays. All ray cells have the same shape. Hawthorn (*Crataegus monogyna*). Small: *Crataegus monogyna*. Upper Franconia, Germany (photo: Aas).

4 MODIFICATION OF THE STEM STRUCTURE

The Primary and Secondary Stages of Growth of Monocotyledons, Macroscopic View

The anatomical structure of most monocotyledons is fairly uniform (4.58-4.68). All of them have closed, collateral vascular bundles, which are embedded in parenchymatous tissue (4.65). The position and proportion of xylem and phloem, and the accompanying sclerenchymatous sheath, are variable. Very long crystals (calcium oxalate rhaphides) are frequent (4.71). The unlignified phloem is clearly visible in photographs taken with transparent light. Stabilization tissue is very obvious in photographs taken with polarized light.

The absence of secondary growth is characteristic of this very large group of species (4.66-4.68). Only few species developed a special mechanism for secondary growth, e.g. *Yucca* and *Dracaena* in the Agavaceae family (4.64, 4.69, 4.70).

4.58 Palm without secondary growth. The slender stem of uniform thickness is characteristic of this tree. Cabbage Palm (Livistonia sp.), Botanical Garden, Puerto de la Cruz, Tenerife.

4.59 Palm without secondary growth that lives in a seasonal climate. Cabbage Palm (Livistonia sp.). Minor periodical growth constraints lead to irregular longitudinal growth per year (arrows). Arizona, USA.

4.60 Coconut Palm (Cocos nucifera) on a windy shore. Stabilization is assured by the cable-like arrangement of very sclerotic, vascular bundles. Salalah, Oman.

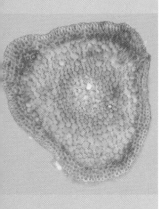

4.61 Dragon Tree (Dracaena draco) on the Canaries.

4.62 Palm stem cross-section with vascular bundles (brown dots), embedded in parenchymatous tissue.

4.63 Soaptree Yucca (Yucca elata) in Arizona, USA.

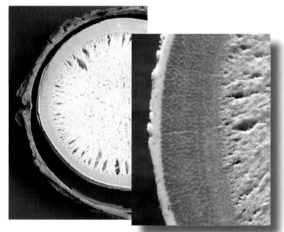
4.64 Stem cross-section of a Dragon Tree (Dracaena draco), Dhofar, Oman. The big, light-coloured center represents the parenchymatous pith. It is surrounded by secondary xylem (yellow ring) and thin periderm. The space between wood and bark was created by shrinkage.

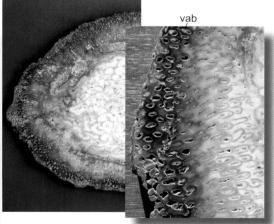
4.65 Stem cross-section of a fossil, Mesozoic palm with vascular bundles (little dots). South Africa.

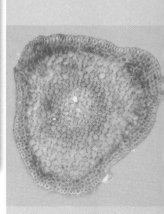

continued next page

4 MODIFICATION OF THE STEM STRUCTURE

The Primary and Secondary Stages of Growth of Monocotyledons, Microscopic View

4 Modification of the Stem Structure

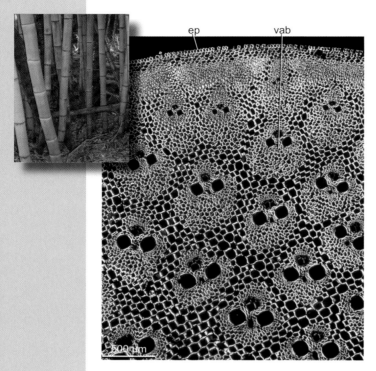

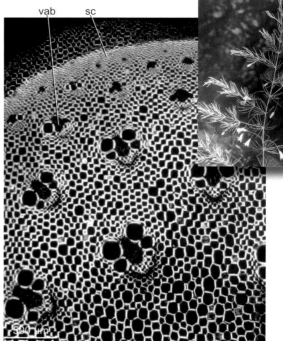

4.66 Cross section of a 3-m-tall monocotyledonous Bamboo (Phyllostachis edulis) with closed, collateral, vascular bundles. There is no secondary growth. Small left: Bamboo in the Botanical Garden of Kyoto, Japan.

4.67 Cross section of an upright monocotyledon shoot with closed, collateral, vascular bundles. There is no secondary growth. Asparagus (Asparagus tenuifolium; polarized light). Small right: from Aeschimann et al. 2004.

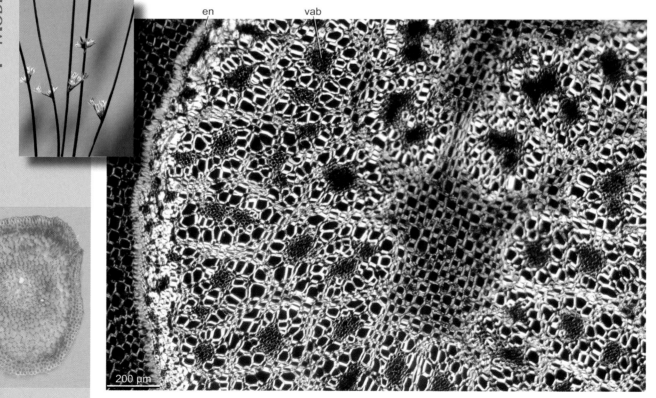

4.68 Cross section of a prostrate monocotyledon rhizome with closed, concentric, vascular bundles. There is no secondary growth. Arctic Rush (Juncus arcticus; polarized light). Small left: from Aeschimann et al. 2004.

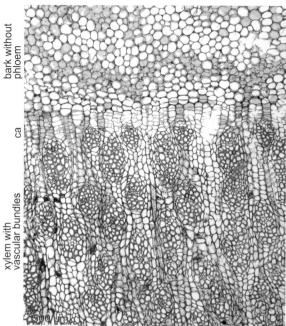

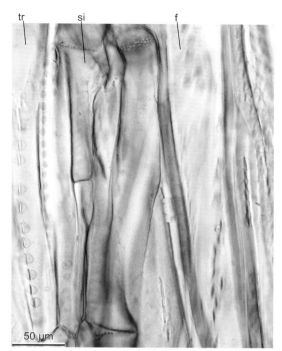

4.69 Transversal section of a monocotyledon with secondary growth. The center of the stem contains central vascular bundles, which consists of xylem and phloem. The bark contains only parenchymatous tissue. There is no phloem. The cambium is between the central part and the bark. Dragon Tree (Dracena serrulata).

4.70 Longitudinal section of sieve tubes with perforated ends in the Dragon tree (Dracena serrulata).

4.71 Elongated crystal bundles (rhaphides) and crystal sand in the parenchyma cells of a Dragon Tree (Dracaena serrulata; polarized light). Small right: Dracaena draco on Tenerife, Canary Islands.

4 MODIFICATION OF THE STEM STRUCTURE

The Secondary Stage of Growth: Conifer Phloem

The structure of conifer phloem is relatively simple. Sieve tubes, parenchyma and fibers are mostly arranged in tangential rows (4.72-4.76). Many cells contain crystals, mostly calcium oxalate, in different forms (4.75). When the girth increases due to secondary growth, the sieve tubes collapse, parenchyma cells enlarge, and local rays dilate. The sclereids are scattered (4.76). The sieve tube areas at the radial sides of phloem cells are characteristic of conifers.

4.72 Cross section of young Silver Fir phloem (Abies alba). The cambium periodically produces several rows of sieve tubes and single rows of parenchyma cells. The parenchymatous rows were formed in spring. Above: Fir bark.

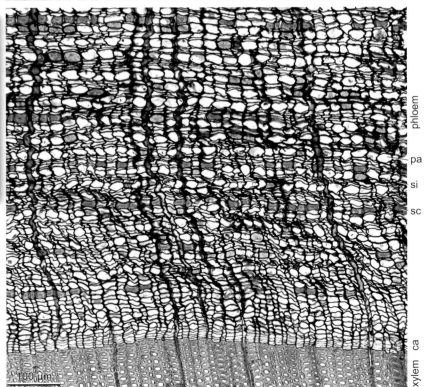

4.73 Cross section of the phloem of a Common Yew (Taxus baccata). The cambium produces sieve cells and parenchymatic cells. Sieve cells collapse after death and the living parenchyma cells expand with age. Above: Yew bark.

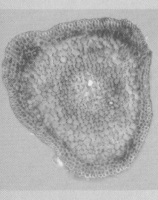

4.74 Cross section of Norway Spruce phloem (Picea abies). Basically, the structure is similar to that of fir, but the sieve tubes have collapsed. Above: Spruce bark.

4.75 Cross section of a Common Yew phloem (Taxus baccata). Tangential rows of sieve tubes with few crystals and crystal sand (little white dots) at their walls.

4.76 Cross section of Savin Juniper phloem (Juniperus sabina). The formation of sieve tubes/parenchyma is regular. This does not represent annual growth zones. Some cells become sclerenchymatous fibers. The resin ducts in the phloem are characteristic of Junipers. Small right: Juniper bark.

4 MODIFICATION OF THE STEM STRUCTURE

87

The Secondary Stage of Growth:
The Phloem of Dicotyledonous Angiosperms

The principle structure is comparable to that of conifers. The cell elements are the same, but there is an enormous variability regarding cell arrangement and size, as well as the non-functional phloem. There is no principal difference between growth forms, such as herbaceous plants, dwarf shrubs, shrubs and trees. Below is a selection of some growth forms from different taxonomic units (4.77-4.85).

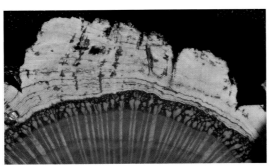

4.77 Macroscopic aspect of the position of the phloem in a stem of a dicotyledonous tree (Cork Oak, Quercus suber). The phloem is located between the xylem and the phellem (cork).

4.78 Macroscopic aspect of a fresh wound in an Ash bark (Fraxinus excelsior). The xylem, phloem and the outer dead bark (rhythidiome) are distinct. The white layer on the xylem is the cambium.

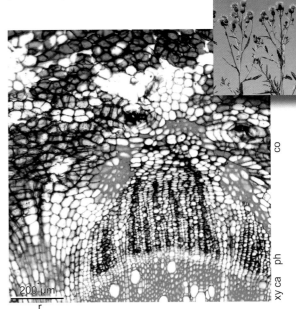

4.79 Annual herbaceous plant shoot, hemicryptophyte. The cambium produces the xylem centripetally, and the phloem centrifugally. The phloem is embedded in the parenchymatous cortex. With increasing girth, the phloem developed cone-like. Creeping Thistle (Cirsium arvense), Asteraceae. Small: from Aeschimann et al. 2004.

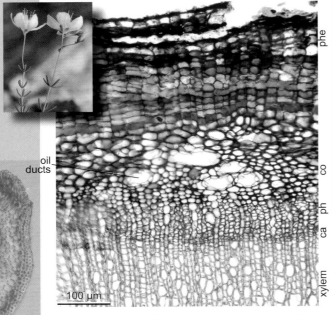

4.80 Dwarf shrub, chamaephyte. On the outside of the almost unstructured phloem is the cortex with oil ducts. Some cork layers (phellem) cover the cortex. Heath-leaved St. Johns-wort (Hypericum coris), Cistaceae.

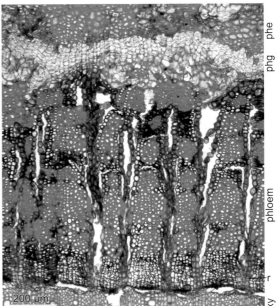

4.81 Small shrub with dense, hard bark. Intensively lignified radial oriented fiber groups are radial orientated between unlignified small rays. German Tamarisk (Myricaria germanica).

4 Modification of the Stem Structure

4.82 Small shrub with a soft, thick bark. The phloem contains some indistinct, probably annual layers. Mezereon (Daphne mezereum), Thymelaeaceae. Small: from Aeschimann et al. 2004.

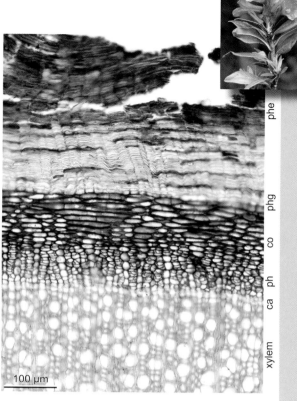

4.83 Dwarf shrub with a dense, wooden stem and thin bark. The cambium produces few phloem cells. Cell expansion is very obvious in the cortex. The multi-layered periderm is very distinct. Rusty-leaved Alpenrose (Rhododendron ferrugineum), Ericaceae. Small: from Aeschimann et al. 2004.

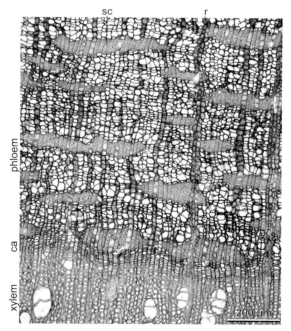

4.84 Tree with dense wood and rough bark. Regular formation of sieve tubes and parenchyma cells (bigger cells are arranged in tangential rows). Hop Hornbeam (Ostrya carpinifolia), Corylaceae.

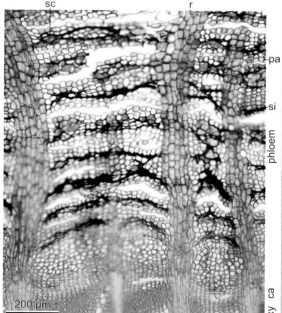

4.85 Little tree with a smooth bark. Regular formation of sieve tubes and parenchyma cells. One year after their formation, the sieve tubes collapsed due to the strength of the periderm belt. Golden Rain (Laburnum anagyroides), Fabaceae.

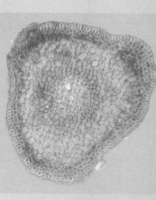

4 MODIFICATION OF THE STEM STRUCTURE

The Secondary Stage of Growth:
Cambial Growth Variants and Successive Cambia

The cambium of most plants continuously produces, during the entire life of the plant, xylem centripetally and phloem centrifugally. That is a „normal" stem. However, many species or genera of some families do not form phloem and xylem continuously, but regularly. At this point, successive cambia should be mentioned. In successive cambia, the lateral primary meristem (cambium) functions only for a limited time-period, then it is replaced by a new cambium, which originated in the parenchyma of the phloem. The new cambium functions in the same way as the primary one, by producing xylem towards the inside and phloem towards the outside of the stem. Some species replace the primary cambium over the entire circumference of the stem, some only locally (4.86–4.91).

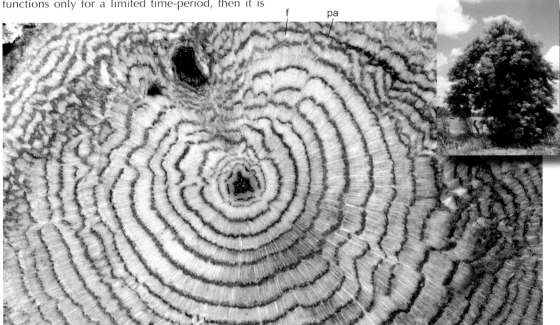

4.86 Stem of the Australian Christmas Tree (Nyutsia floribunda), Loranthaceae, with a „successive cambium". The distinct „tree rings" actually reflect the periodic production of parenchyma, vessels and fibers. This formation is not climatically-induced. Small right: Christmas tree in Western Australia.

4.87 Australian liana stem with alternating xylem-phloem compartments. Small left: twisted liana stem in Queensland, Australia.

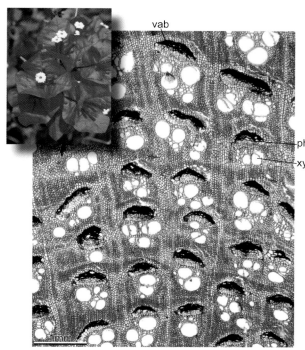
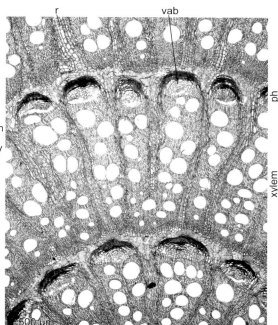

4.88 Cross-section of a liana-like Bougainvillea stem (Bougainvillea spectabilis), Nyctaginaceae. Large vascular bundles with big vessels in the xylem, and mostly collapsed sieve tubes in the phloem, are characteristic of this plant. The bundles are surrounded by dense, lignified fiber tissue and indistinct rays. Small left: Bogainvillea spectabilis flowers with red spathaceous bracts.

4.89 Caper bush stem cross-section (Capparis sp.), Capparidaceae. The cambium periodically produces xylem/phloem zones. The phloem remains functional only in the immediate neighbourhood of the xylem, later it collapses.

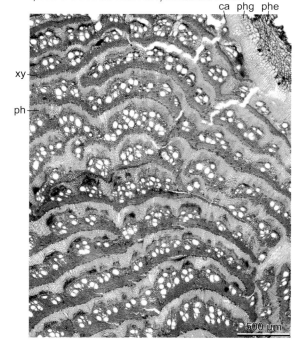

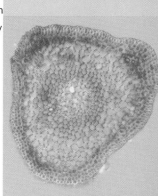

4.90 Cross-section of an annual Goosefoot shoot (Chenopodium sp.), Chenopodiaceae. The stem-internal parenchyma and phloem and the fiber-vessel bands are characteristic of a successive cambium. The most active cambium is just underneath the bark.

4.91 Cross-section of an annual White Goosefoot root (Chenopodium album), Chenopodiaceae. The stem-internal parenchyma and the fiber-vessel bands are characteristic of a successive cambium. The xylem and the spots of phloem, which are surrounded by thin parenchyma cells, are well-differentiated.

4 MODIFICATION OF THE STEM STRUCTURE

THE THIRD STAGE OF GROWTH: THE PERIDERM

As plants become older, the epidermis is replaced by another protective tissue: the periderm (4.92-4.94). A secondary cambium, the phellogen, produces mainly centrifugal, tangential, flat cells with suberized walls (4.95-4.98). Their life expectancy is very short, and the dead cells contain air: this is the phellem. Towards the inside, a few parenchyma cells (phelloderm) grow.

As the circumference increases, the first phellogen is replaced by a new one, somewhere in the outer part of the live phloem (4.99). As a consequence, some older parts of the phloem die and finally drop off the stem. This dead part outside the cambium is mostly called bark. All conifers and dicotyledons and some monocotyledons (4.97), including annuals (4.100) and perennials of all sizes, produce a periderm. Its anatomical structure may differ greatly.

Top right: 4.92 Casuarina dendata stem disk (Casuarinaceae). The cog-wheel-like periphery consists, in fact, of phelloderm layers. The compact, dark belt below is the phloem.

Middle right: 4.93 Saw Banksia (Banksia serrata) stem disk (Proteaceae). The internal structure of the periderm is irregular. The light spots are groups of sclereids.

Bottom right: 4.94 Two- to three-year-old Casuarina branch (Casuarinaceae). More phellem than xylem is produced. The layers within the phellem reflect regular, intra-annual phellogen activity.

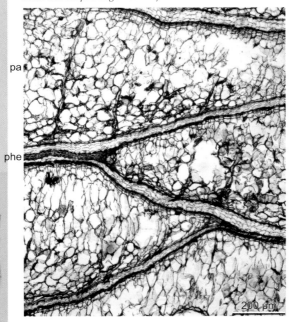

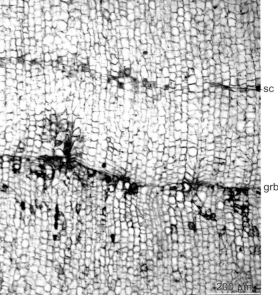

4.95 Six-year-old Norway Spruce top shoot (Picea abies). The phellogen has been replaced several times and has produced many irregular layers of phellem.

4.96 Cork of a Cork Oak (Quercus suber). It consists mainly of thin, suberinized cell walls. A few cells are sclerenchymatic. This kind of tissue was observed for the first time by Robert Hooke in 1665.

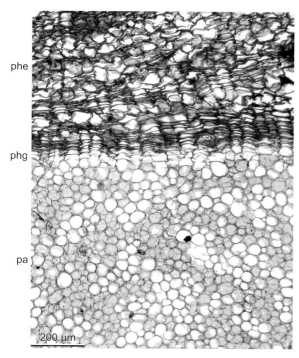

4.97 Periderm of a monocotyledon *Yucca* sp. with secondary, radial growth. The phellogen produced a periderm (brown), like a dicotyledon, but there is no phloem on the outside of the cambium. The blue part is pure parenchymatous tissue.

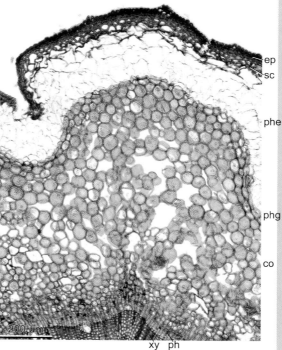

4.98 One-year-old Norway Spruce top shoot (*Picea abies*). The cortex is covered by phellem, very thin-walled cork cells and the former epidermis. The cork cells are a product of the secondary meristem (phellogen).

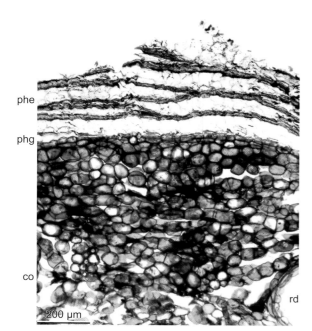

4.99 Periderm of an old Stone Pine stem (*Pinus cembra*). The cork layers separate old parts of the phloem.

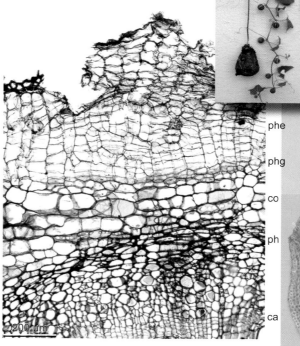

4.100 Periderm of an annual, liana-like, White Bryony shoot (*Bryonia dioica*). The peripheric cork layer was produced by the phellogen: that is the unicellular layer just underneath the cork. Small right: *Bryonia verrucosa* with a root tuber. Tenerife.

4 Modification of the Stem Structure

Chapter 5

Modification of the Xylem Within a Plant

The large internal physiological and mechanical variation within plants results in their large anatomical variability.

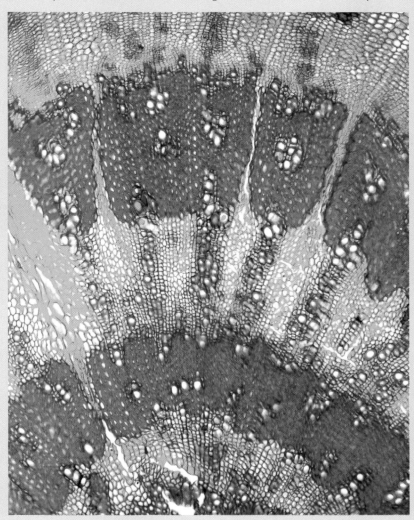

Cross section of a stem of a hemicryptophytic herb, Cresse (Rorippa stylosa).

Modification of the Xylem Within a Plant
Conifer: Root, Twig and Stem

General features of the xylem structure are species-specific, many attributes are modified by the growth environment.

The wood anatomy of **branches and twigs** reflects changing physiological conditions and mechanical forces (5.1, 5.4, 5.8). Their relatively small vessels - in relation to the stem - indicate some physical tension. Genetic differences between the taxa are easily recognizable in those branches and twigs which are not subject to much mechanical stress.

Stem wood generally represents mechanically stable conditions (5.2, 5.5, 5.9). Therefore, genetic differences between taxa are best expressed in stem wood. Its anatomy is mainly influenced by the amount of water flow from the roots to the crown, and by major mechanical forces.

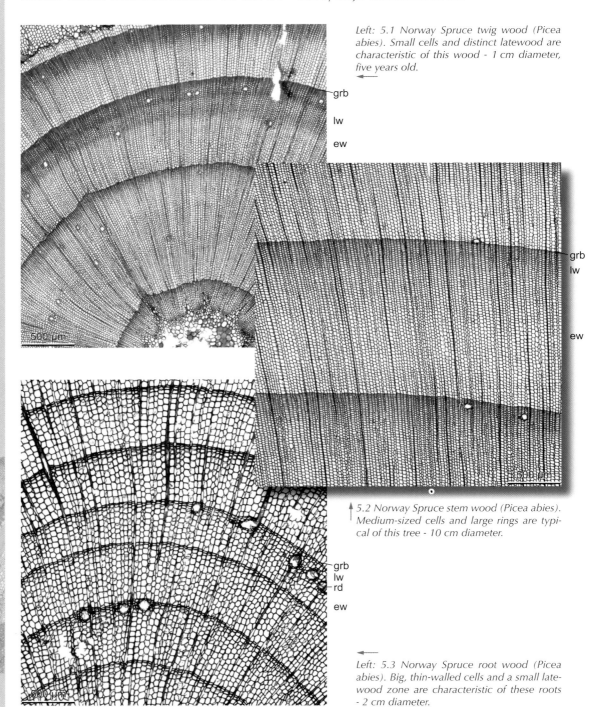

Left: 5.1 Norway Spruce twig wood (Picea abies). Small cells and distinct latewood are characteristic of this wood - 1 cm diameter, five years old.

5.2 Norway Spruce stem wood (Picea abies). Medium-sized cells and large rings are typical of this tree - 10 cm diameter.

Left: 5.3 Norway Spruce root wood (Picea abies). Big, thin-walled cells and a small latewood zone are characteristic of these roots - 2 cm diameter.

Deciduous Tree: Root, Twig and Stem

Root wood is subject to great anatomical variability because of extremely variable soil conditions (5.3, 5.6, 5.7). Genetic differences between taxa are only recognizable, if at all, in the cell wall structure. Root wood is mainly influenced by the amount of water to be transported and by very variable forces of tension.

Within individuals, the wood-anatomical differences between stem, roots and twigs reflect the environmental conditions. When roots are exposed, they adopt a stem structure, whereas twigs buried in the ground take on root structure (5.10-5.14).

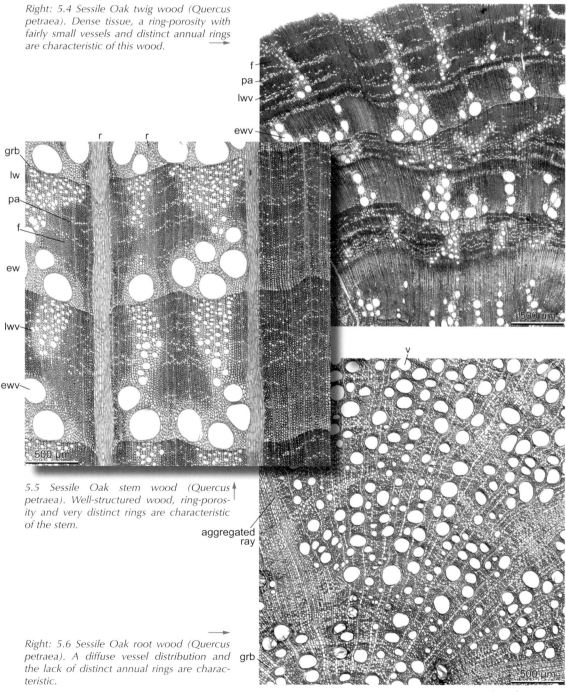

Right: 5.4 Sessile Oak twig wood (Quercus petraea). Dense tissue, a ring-porosity with fairly small vessels and distinct annual rings are characteristic of this wood.

5.5 Sessile Oak stem wood (Quercus petraea). Well-structured wood, ring-porosity and very distinct rings are characteristic of the stem.

Right: 5.6 Sessile Oak root wood (Quercus petraea). A diffuse vessel distribution and the lack of distinct annual rings are characteristic.

continued next page

5 Modification of the Xylem Within a Plant

Modification of the Xylem Within a Plant
Deciduous Tree: Root, Twig and Stem

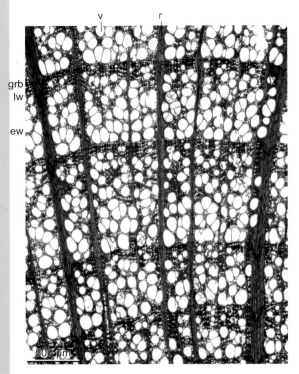

5.7 Common Beech root wood (Fagus sylvatica) with a large number of big vessels.

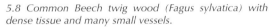

5.8 Common Beech twig wood (Fagus sylvatica) with dense tissue and many small vessels.

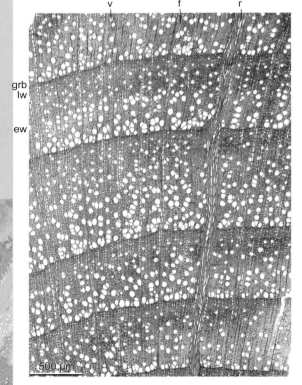

5.9 Common Beech stem wood (Fagus sylvatica) with dense tissue, semi-ring-porosity and very distinct annual rings.

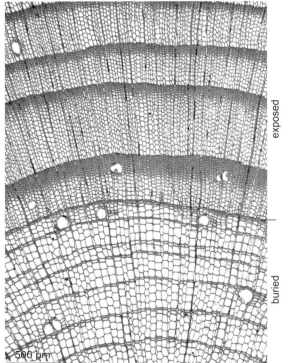

5.10 Exposed Mountain Pine root (Pinus mugo). Large, thin-walled cells are characteristic of the root structure.

FROM ROOT TO STEM STRUCTURE

5.11 Exposed adventitious Cottonwood tree roots (Populus sp.). The stem was buried by flood sediment. Under the dark and wet conditions, a new root system was initiated. Many years later, this particular part was exposed by erosion.

5.12 Exposed Common Beech roots (Fagus sylvatica). Externally, they look like roots, but the xylem produced in the root phase is different from that formed when the roots were not exposed. Beech forest, Zürich, Switzerland.

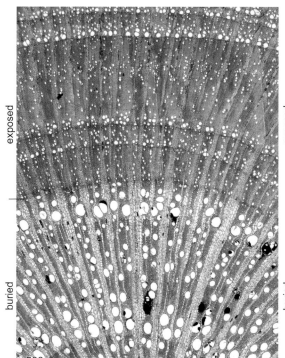

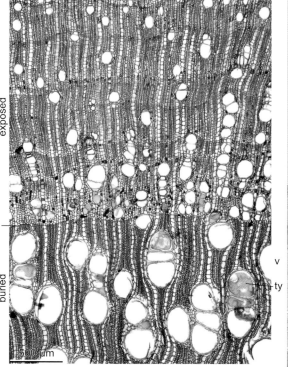

5.13 Root/stem wood of an exposed Peach Tree root (Prunus persica). The ring-porosity with large earlywood vessels is characteristic of the root structure, while dense fiber tissue and small earlywood vessels are typical of the stem structure.

5.14 Lotus Tree root/stem wood (Ziziphus lotus) of an exposed root, which has grown in a riverbed in the arid climate of Oman. Characteristic of the root wood are very big vessels and the absence of annual rings, whilst small vessels and annual rings are typical of the stem wood.

5 MODIFICATION OF THE XYLEM WITHIN A PLANT

Modification by Aging: Changing Growth Forms

Morphological changes of tree and leaf shapes reflect the aging process. Tree growth forms emphasize the phenomena of tree aging, for example, a leading, dominant shoot is typical of young conifers (Roloff 2001) while old trees show a flat crown, e.g. in fir, cedar, araucaria and others (5.15-5.18). Wood anatomical characteristics may differ with age (5.19, 5.20). In a few genera, such as in *Ilex* and *Hedera*, leaf shape varies, depending on the shoot's age (5.21, 5.22).

5.15 A young Monkey-puzzle Tree (*Araucaria araucana*). The top shoot grows vertically. Lonquimai, Patagonia, Argentina.

5.16 An old Monkey-puzzle Tree (*Araucaria araucana*). The top of the crown is flat. Lonquimai, Patagonia, Argentina.

5.17 A young, fast grown Spanish Fir (*Abies pinsapo*). The top shoot grows vertically. The tree has not reached the final height. Arboretum Birmensdorf, Switzerland.

5.18 Old, slow growing Spanish Firs (*Abies pinsapo*). The crown tops are flat and don't grow higher anymore. Sierra da Ronda, Spain.

Changing Growth and Leaf Forms

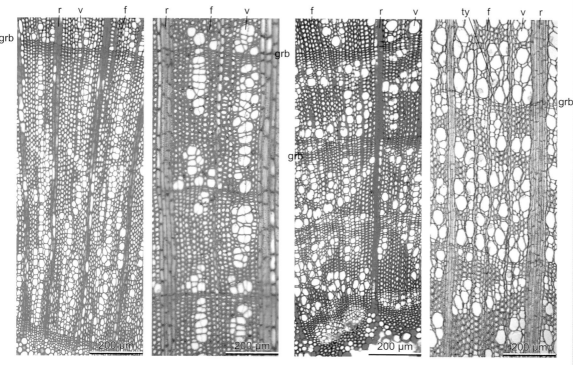

5.19 Cross-sections of juvenile (left) and adult (right) parts of the xylem. Common Holly (Ilex aquifolium).

5.20 Cross-sections of juvenile (left) and adult (right) parts of the xylem. Common Ivy (Hedera helix).

5.21 Top: Spiny juvenile leaves. Below: Adult leaves with a smooth leaf margin. Common Holly (Ilex aquifolium).

5.22 Top: Lobed juvenile leaves. Below: Elliptical adult leaves of the Common Ivy (Hedera helix).

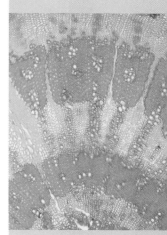

5 Modification of the Xylem Within a Plant

Modification by Aging:
Changing Wood Anatomical Structures

Quantitative (reversible) physiological aging is expressed in the slowing down of cell division, changing cell differentiation, and the formation of secondary and tertiary walls. Qualitative (irreversible) aging causes a loss of protoplasts, complete lignification and the loss of cell division capacity (Bosshard 1982).

As these process-oriented considerations generally are not reflected in the anatomical structure, biologists use the terms „juvenile wood" and „adult wood" (5.23). Juvenile wood is produced during that period when the cell dimension progressively increases, whereas adult wood is characterized by cells of more or less constant size.

Juvenile wood cells and rays are usually smaller than those of adult wood (5.24-5.27), but the position of young and old tissue within the plant can modify this tendency in an unpredictable manner. In a few species, libriform fiber groups in the latewood seem to indicate juvenile wood (5.28, 5.29).

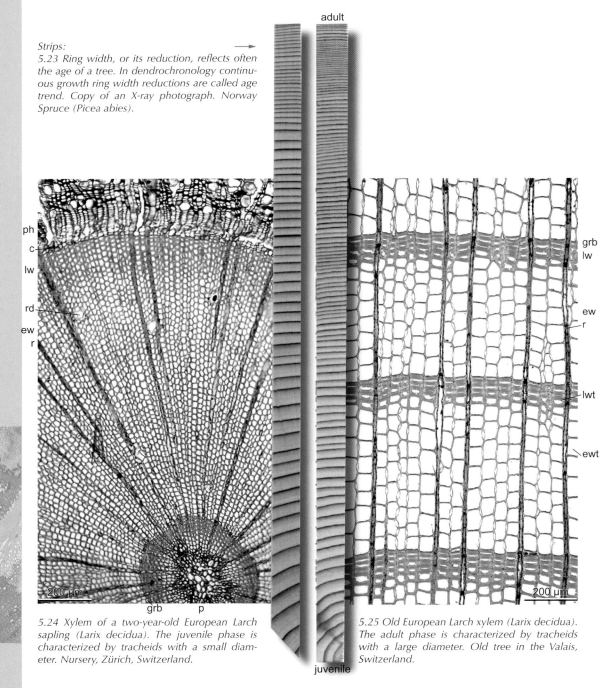

Strips:
5.23 Ring width, or its reduction, reflects often the age of a tree. In dendrochronology continuous growth ring width reductions are called age trend. Copy of an X-ray photograph. Norway Spruce (*Picea abies*).

5.24 Xylem of a two-year-old European Larch sapling (*Larix decidua*). The juvenile phase is characterized by tracheids with a small diameter. Nursery, Zürich, Switzerland.

5.25 Old European Larch xylem (*Larix decidua*). The adult phase is characterized by tracheids with a large diameter. Old tree in the Valais, Switzerland.

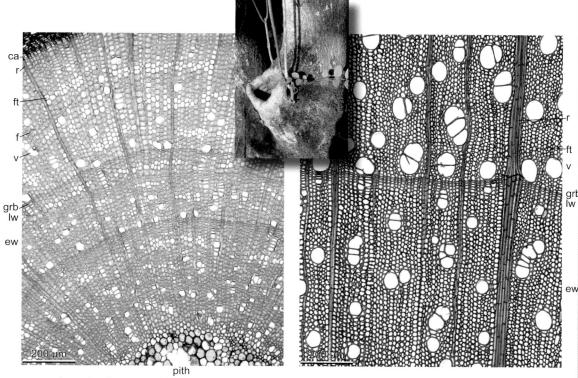

5.26 Xylem of a three-year-old Sycamore sapling (Acer pseudoplatanus). The juvenile phase is characterized by small vessels and rays. Center top: Long shoots around a scar on an Ash stem (Fraxinus excelsior). The xylem in the long shoot is in a juvenile, the one in the stem in an adult stage.

5.27 Old Sycamore xylem (Acer pseudoplatanus). The adult phase is characterized by large vessels and rays.

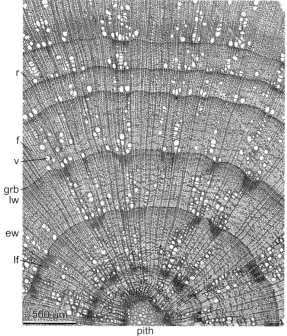

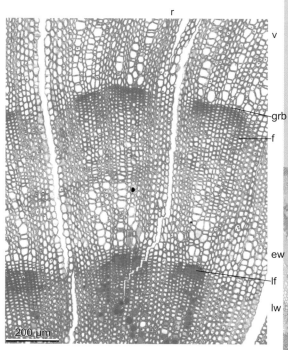

5.28 Xylem of a suppressed Hornbeam (Carpinus betulus). The juvenile phase in the stem's center is characterized by distinct groups of libriform fibers at the tree-ring boundaries. Such groups disappear with progressive aging.

5.29 Groups of thick-walled libriform fibers in a juvenile stage of the herbaceous plant Green Hellebore (Helleborus viridis).

5 MODIFICATION OF THE XYLEM WITHIN A PLANT

Modification by Aging:
Change of Phloem and Periderm Structures

As the stem expands, the aspect of its bark changes: it is smooth when the tree is young, rough when it gets older (5.30-5.32).

The anatomical changes are dramatic: the epidermis is replaced by one or several layers of periderm, whilst the cortex and the phloem are locally inactivated and compressed (5.33-5.37). Finally, all these parts are physiologically isolated by the formation of a phellogen (cork cambium).

5.31 Smooth bark of a young Apple Tree (Malus sylvestris, old long shoot of 5.32). Würzburg, Germany.

5.30 Old stem with a rough bark and a young adventitious shoot with a smooth, green bark. Honey Mesquite (Prosopis glandulosa). Tucson, Arizona, USA.

5.32 Rough bark of an old Apple Tree (Malus sylvestris). Würzburg, Germany.

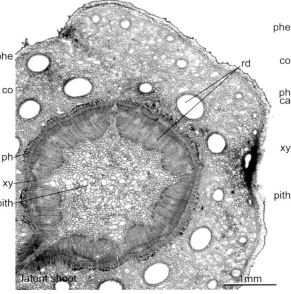

5.33 One-year-old top shoot of a Stone Pine (Pinus cembra). The cortex is very large in relation to the xylem. The phloem between xylem and cortex is very small.

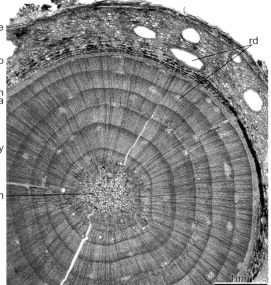

5.34 Six-year-old top shoot of a Stone Pine (Pinus cembra). The xylem is much bigger than that of the younger shoot above; the phloem is a little bit larger, and the cortex has become compressed, as indicated by the oval resin ducts.

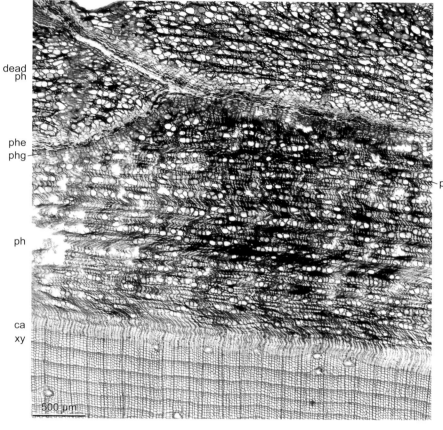

5.35 Xylem and bark of an old Stone Pine (Pinus cembra). The xylem and phloem have become much bigger, and the cortex has been replaced by periderm.

5 MODIFICATION OF THE XYLEM WITHIN A PLANT

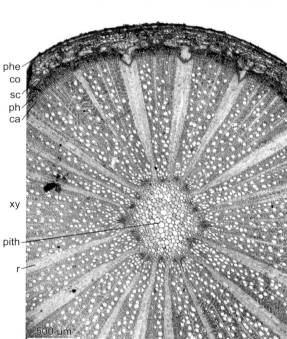

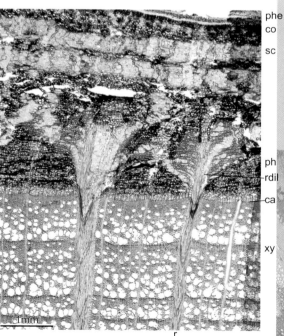

5.36 One-year-old Common Beech seedling (Fagus sylvatica). The big xylem is surrounded by a small cortex and a very small phloem.

5.37 Xylem and bark of an old Beech (Fagus sylvatica). The phloem has expanded due to the dilatation of large sclerotic rays and the expansion of outer sclerotic cells.

Chapter 6

Modification of the Xylem and Phloem by Ecological Factors

This chapter illustrates important anatomical features which can be interpreted as reactions to environmental stress.

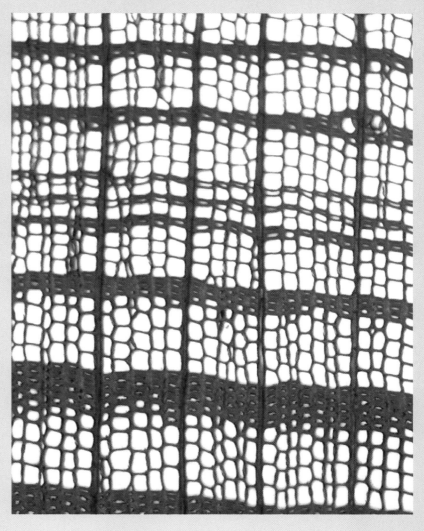

Collapsed vessels in a little shrub stem (Salix glaucosericea).

Intra-Annual Density Fluctuations, Phenolic and Crystal Deposits

Intra-annual density fluctuations indicate short-term events during the growing season. They mainly occur in the xylem of perennial plants in temperate, arid and tropical climates and are virtually absent in cold climates.

Zones of small, thick-walled tracheids or fibers in the earlywood and latewood are mainly found in conifers, but they also occur in the xylem of broadleaf trees (6.2-6.4).

Genetically induced, regular, tangential parenchyma (6.6, 6.7), vessel or fiber (6.5, 6.8) and phloem bands (6.1) cannot be interpreted ecologically.

Intra-annual phenolic (6.9) and crystal deposits in the xylem are rare, whereas crystal deposits in the phloem (6.10) are quite frequent.

6.1 Intra annual growth fluctuations in a stem of a Salt Tree (Haloxylon persicum, Chenopodiaceae). The tree was planted 14 years before felling. The stem has no distinct annual rings but shows many alternating tangential xylem/phloem bands. The white lines indicate probable annual growth rings. Small left: Haloxylon sp., Republic of Mongolia (photo: W. Schulze).

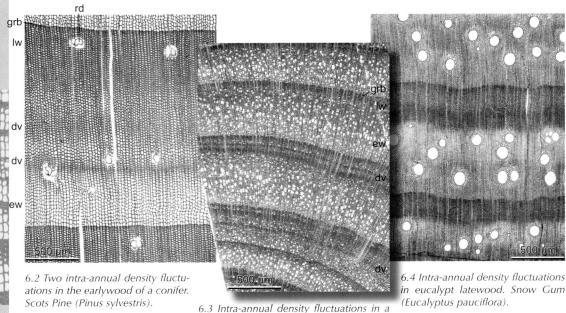

6.2 Two intra-annual density fluctuations in the earlywood of a conifer. Scots Pine (Pinus sylvestris).

6.3 Intra-annual density fluctuations in a small broadleaf shrub. Dusky Foxglove (Digitalis obscura).

6.4 Intra-annual density fluctuations in eucalypt latewood. Snow Gum (Eucalyptus pauciflora).

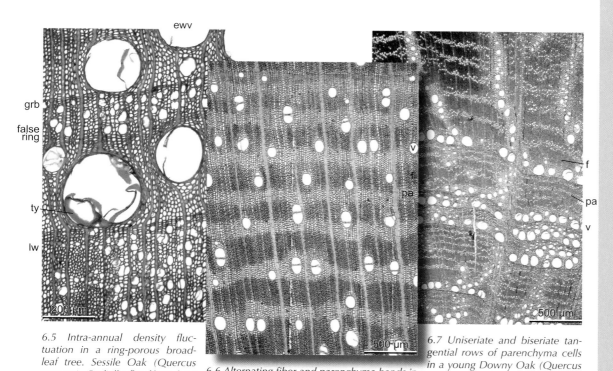

6.5 Intra-annual density fluctuation in a ring-porous broadleaf tree. Sessile Oak (*Quercus petraea*). Radially flat fibers limit earlywood and latewood.

6.6 Alternating fiber and parenchyma bands in a Fig stem (*Ficus carica*).

6.7 Uniseriate and biseriate tangential rows of parenchyma cells in a young Downy Oak (*Quercus pubescens*).

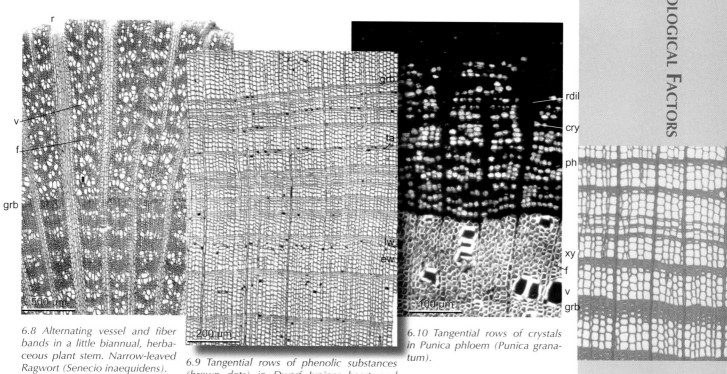

6.8 Alternating vessel and fiber bands in a little biannual, herbaceous plant stem. Narrow-leaved Ragwort (*Senecio inaequidens*).

6.9 Tangential rows of phenolic substances (brown dots) in Dwarf Juniper heartwood (*Juniperus nana*).

6.10 Tangential rows of crystals in Punica phloem (*Punica granatum*).

6 MODIFICATION OF XYLEM AND PHLOEM BY ECOLOGICAL FACTORS

Intra-Annual Cell Collapse, Callous Tissue and Ducts

Irregularities in the xylem are mainly induced by extreme weather conditions, such as frost and drought, and by mechanical stress. Loss of pressure in thin-walled cells causes cell collapses (6.11). After such events, callous cells repair the damage (6.14-6.17). Collapsed and callous cells are typical of so-called frost rings (6.12, 6.13). Physiological and mechanical stress triggers the formation of tangential rows of resin ducts in conifers, and of gum ducts in broadleaf trees (6.16-6.19).

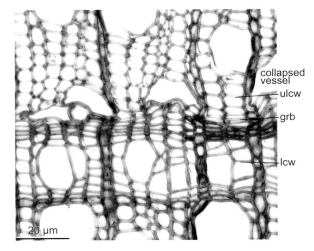

Right: 6.11 Collapsed vessels in a little shrub stem. Willow (*Salix glaucosericea*).

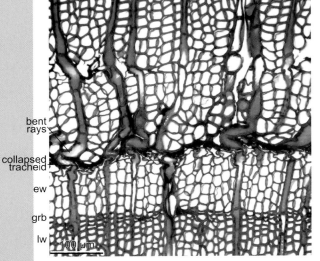

6.12 Collapsed tracheids in the earlywood of a conifer as a result of late frost. Siberian Larch (*Larix sibirica*).

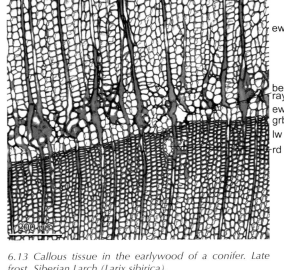

6.13 Callous tissue in the earlywood of a conifer. Late frost. Siberian Larch (*Larix sibirica*).

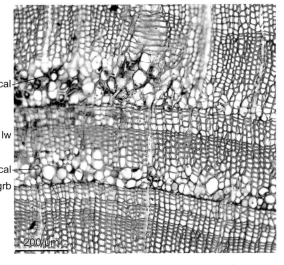

6.14 Callous tissue in the earlywood of a conifer. Injury. Mountain Pine (*Pinus mugo*).

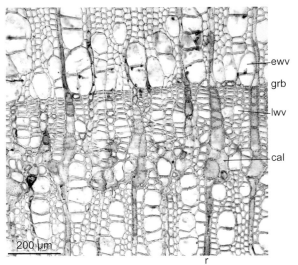

6.15 Callous tissue in the latewood of a broadleaf tree. Frost ring after a volcanic eruption. *Nothofagus pumilio*.

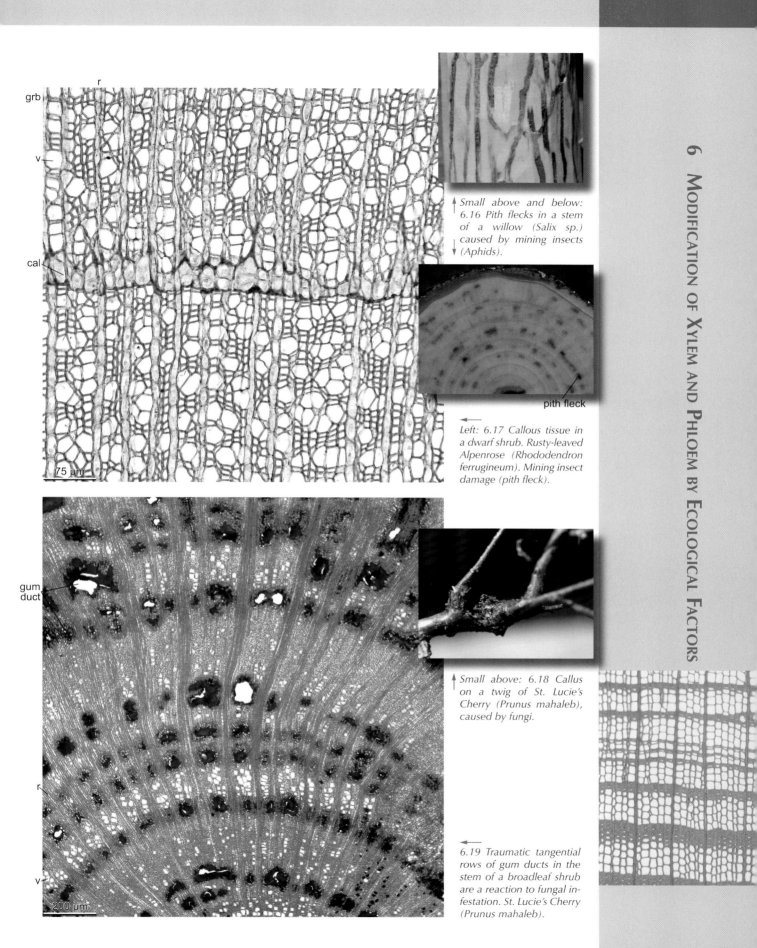

Small above and below: 6.16 Pith flecks in a stem of a willow (Salix sp.) caused by mining insects (Aphids).

Left: 6.17 Callous tissue in a dwarf shrub. Rusty-leaved Alpenrose (Rhododendron ferrugineum). Mining insect damage (pith fleck).

Small above: 6.18 Callus on a twig of St. Lucie's Cherry (Prunus mahaleb), caused by fungi.

6.19 Traumatic tangential rows of gum ducts in the stem of a broadleaf shrub are a reaction to fungal infestation. St. Lucie's Cherry (Prunus mahaleb).

6 Modification of Xylem and Phloem by Ecological Factors

Interannual Variation of Latewood Zones

Climatological interpretations of X-ray densitometric studies are mainly based on latewood density variations. Latewood density is a composite value of the number, size and cell wall thickness of tracheids (6.20). High latewood densities in conifers from cool, moist regions reflect warm, while low latewood densities reflect cool summers (see Introduction). However, they also correlate with short and long growing seasons. The so-called "light rings" (Filion *et al.* 1986) also include rings that have a low latewood density. Latewood formation is very much reduced after insect attacks (Schweingruber 2001). The latewood structure of broadleaf trees also varies, but might reflect growing season length (Skomarkova *et al.* in prep.).

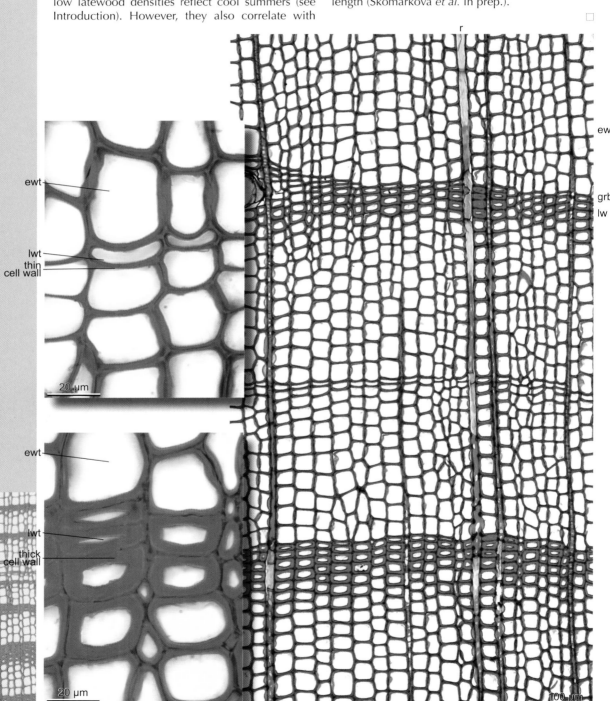

6.20 Tree ring sequence with wide and narrow latewood. Top left: Latewood with thin-walled tracheids. Bottom left: Latewood with thick-walled tracheids. Scots Pine (Pinus sylvestris).

Chapter 7

Modification of Organs

Plant reproduction and occupation of ecological niches are only possible by the modification of shoots.

Rose spine (Rosa sp.)

Modification of Shoots:
Long and Short Shoots

The formation of plant crowns depends on the initiation of buds and the variable growth of twigs (Roloff 2001). Some twigs grow fast and turn into long shoots; others grow slowly and become short shoots (7.1-7.5). Genetic and ecological factors determine the growth of twigs. The genetic influence is very obvious in some conifers, such as larch (7.7-7.10). Annual longitudinal growth may be determined by bud-scale scars. Annual radial xylem growth is indicated by the ring width (7.6). □

7.1 Short shoots on a long One-seed Hawthorn shoot (Crataegus monogyna).

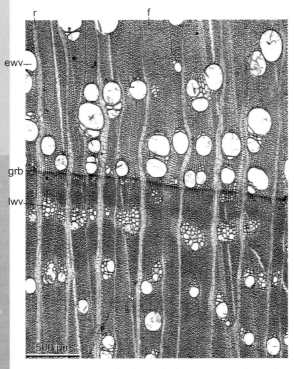

7.2 Cross-section of a long Black Locust tree shoot (Robinia pseudoacacia). In the large rings, the latewood is wide and dense.

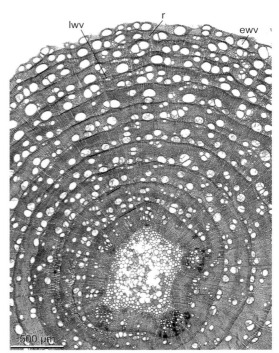

7.3 Cross-section of a short Black Locust tree shoot (Robinia pseudoacacia). In the very small rings, the latewood is almost absent.

7 MODIFICATION OF ORGANS

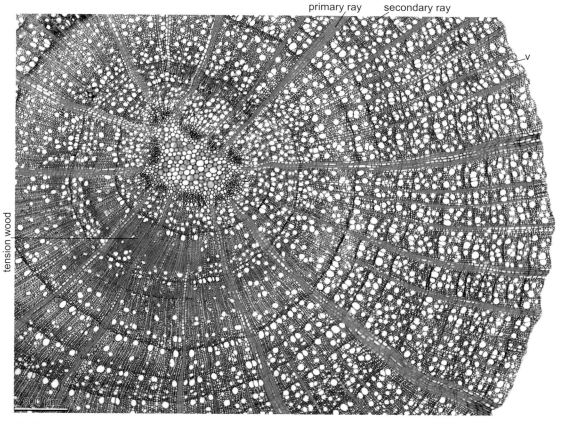

7.4 Cross-section of a short shoot of Beech (*Fagus sylvatica*). The narrow rings are typical of this short shoot. The first wide rings indicate that the twig started as a long shoot.

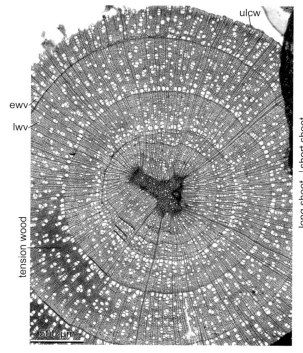

7.5 Cross-section of a long shoot of Silver Birch (*Betula pendula*) with large rings.

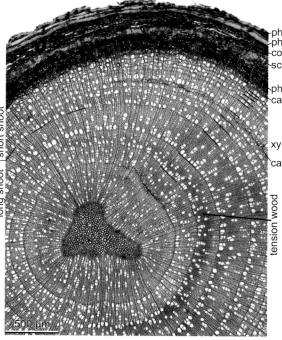

7.6 The transition from a long-shoot-phase to a short-shoot-phase in Silver Birch (*Betula pendula*) is reflected in a sudden radial growth change.

continued next page

Modification of Shoots:
Long and Short Shoots

7.7 Short shoots and male flowers on a horizontal, long shoot of European Larch (Larix decidua).

7.8 Cross-section of a short shoot of a European Larch tree (Larix decidua). The bark is extremely thick in relation to the xylem. The 18-year-old short shoot does not contain any rings in the xylem.

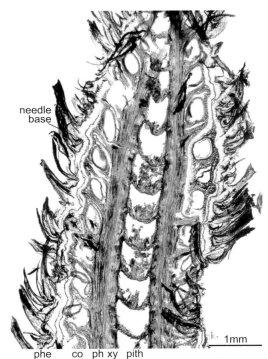

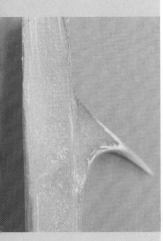

7.9 Longitudinal section of a short shoot of European Larch (Larix decidua). The chambers in the pith represent annual shoots.

7.10 Microscopic section of a twig of Common Beech (Fagus sylvatica). The very small rings in the center indicate that the twig remained in the short-shoot-phase for many years. Due to suddenly improved light conditions, after a neighbouring tree had been felled, the twig went into a long-shoot-phase. This is shown by the large rings towards the periphery.

Modification of Shoots:
Shedding Needles, Male and Female Flowers

After a certain number of years, evergreen conifers shed their needles, male flowers (7.11) and cones. At first, these parts are isolated from the twig, then dropped and, finally, the wound becomes overgrown. The exact position of this wound is datable dendrochronologically. On the basis of that observation, Jalkanen *et al.* (1995) were able to date the life expectancy of needles over a time-period of more than 100 years.

Programmed cell death determines when the needles are shed. They break off at an anatomically fixed position (the needle base), after this tissue has dried. The life expectancy of pine needles varies from 3 to 15 years (7.12). Male flowers are shed after a few weeks (7.13), whereas the cones remain on the twig for many years (7.14).

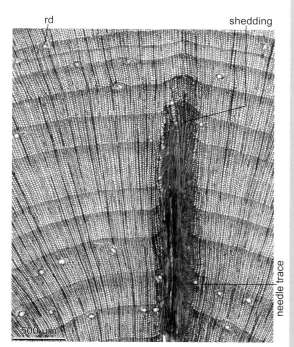

7.11 Spirally arranged traces of male flowers on a long shoot of Mountain Pine (*Pinus mugo*).

7.12 Needle trace on a twig of Mountain Pine (*Pinus mugo*). The needle was alive for over eight years. After shedding of the needle, callus formed for two years. For a few years, the overgrown needle trace may still be identified as a bent latewood zone.

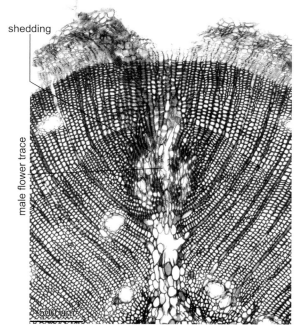

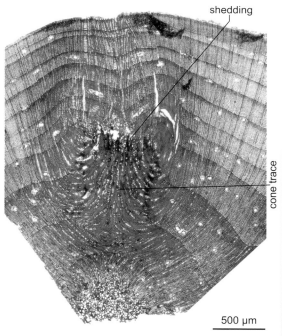

7.13 Overgrown trace of a male flower of Mountain Pine (*Pinus mugo*). This flower had originated in the pith; it was a latent shoot. The twig shed the flower during the first year's latewood formation. Callus was formed until the following year. The open space in the periderm indicates where the flower was shed.

7.14 Overgrown trace of a female cone of Mountain Pine (*Pinus mugo*). It remained on the twig for four years and was shed during earlywood formation. The two following years are characterized by callus formation. It took several years until the differentiation process normalized again.

Shedding of twigs see Abscission pp. 64.

Modification of Shoots:
Thorns and Spines

A very effective method for a plant to resist grazing is the formation of spins and thorn. The evolutionary pressure of herbivores was so strong that stems, branches, twigs and leaves developed thorns (7.15-7.20), and the bark developed spines (7.21, 7.22). All defence mechanisms are based on extreme cell wall growth, the lignification of fibers and parenchyma cells (7.17), and the formation of a sharply pointed tip. Thorns are metamorphosed short shoots or leaf veins (7.19, 7.20), for example in *Berberis*. Spines form from cortex cells, for example in rose twigs (7.21, 7.22).

7.15 Transformation from a short shoot (flowering part) to a thorn (tip). Blackthorn (Prunus spinosa).

7.16 Honey Locust (Gleditsia triacanthos). A shoot transformed into a thorn.

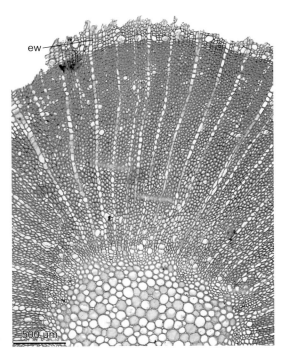

7.17 Cross section of a thorn of Blackthorn (Prunus spinosa) with two leaves. Characteristic is the pith and the absence of vessels. The fibers are very thick-walled.

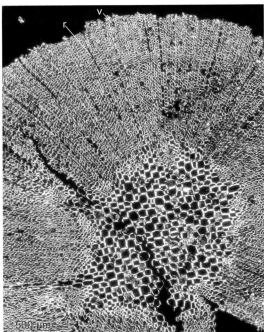

7.18 Cross section of a thorn on a one year-old long shoot of an Apple Tree (Malus sylvestris). Characteristic is the pith and the presence of very small vessels (polarized light).

7.19 Barberry thorns (Berberis sp.). The thorns are relicts of leaves.

7.20 Cross section of a thorn of a cultivated Barberry (Berberis sp.). Characteristic are the vascular bundles in the center which are surrounded by a thick walled parenchymatic tissue.

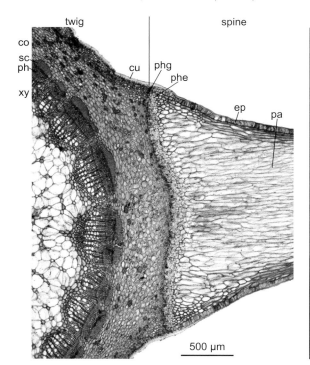

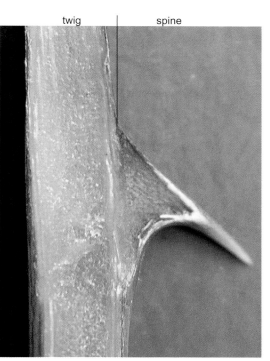

7.21 Cross section of a rose twig and a longitudinal section of a spine. The spine is a product of a phellogen and consists of parenchyma cells. The spine is connected to the twig by a cork layer (phellem). Field Rose (Rosa arvensis).

7.22 Macroscopic view of a Rose spine (Rosa sp.).

7 MODIFICATION OF ORGANS

123

Modification of Shoots:
Vertical, Horizontal and Drooping Twigs

The anatomical shoot structure is influenced by the shoot's position on the tree (7.23). The eccentric growth of horizontal twigs is notable (7.24, 7.25). The lower side of conifers is stabilized by compression wood and the upper side of broadleaf trees by tension wood. Rings are often absent on the upper side (7.24, 7.25). Lack of light frequently causes early death (7.28, 7.29).

Small and often absent rings are characteristic of drooping twigs. Xylem growth is minimal. The number of growth rings is often lower than the number of annual shoots (7.23, 7.26, 7.27).

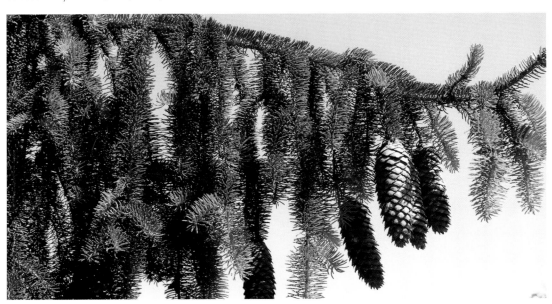

7.23 Norway Spruce branch (Picea abies) with hanging twigs.

7.24 Lower side of an eccentric Norway Spruce branch (Picea abies) with dense compression wood.

7.25 Upper side of the same eccentric branch of Norway Spruce (Picea abies). There are small rings with distinct latewood zones, but without compression wood.

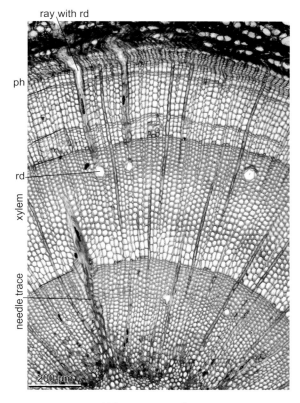

7.26 Five-year-old hanging twig of Norway Spruce (Picea abies). Since mechanical forces are small, the last rings are very small and the tracheids thin-walled.

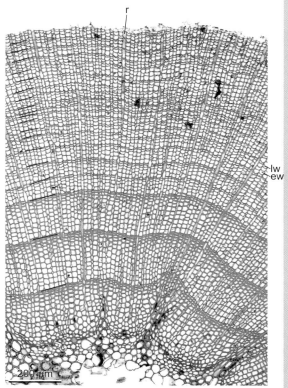

7.27 Twenty-five-year-old hanging twig of Norway Spruce with 20 rings (Picea abies); the age was determined by counting the bud-scale scars.

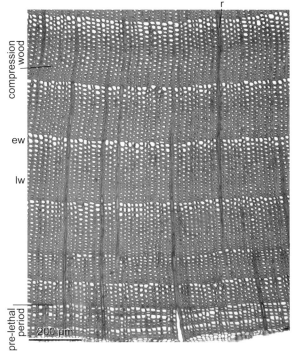

7.28 Lower side of a dead branch of Norway Spruce (Picea abies) in a plantation. Three years before its death, a sudden growth reduction indicated a reduction in vitality.

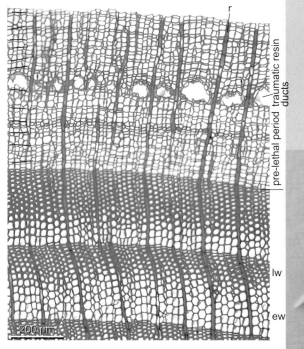

7.29 Dead branch of an old Stone Pine (Pinus cembra) at the upper timberline. Many years before the branch died, a sudden growth reduction indicates a reduced vitality. The rings are no longer clearly delimited.

7 MODIFICATION OF ORGANS

Modification of Shoots:
Latent and Adventitious Shoots

Plant morphology (telome) is mainly a product of the formation of latent and adventitious shoots. Latent shoots originate from the pith (7.30, 7.31). Adventitious shoots are a product of lateral meristems (cambium) or parenchyma cells (7.32-7.37).

Environmental stress, for example the loss of the stem or crown, or leaf damage caused by insect attack or frost, gives rise to the formation of adventitious shoots. Stems covered by soil trigger the formation of secondary roots (7.37).

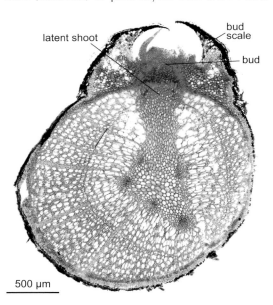

7.30 Bog Whortleberry twig (Vaccinium uliginosum) with a latent shoot. The shoot originated in the pith.

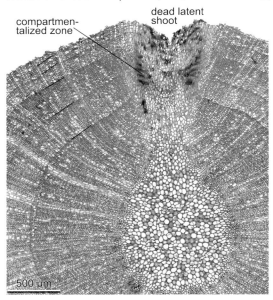

7.31 Buried Common Ivy twig (Ilex aquifolium) with a latent shoot. The break-off point is isolated from the functional tissue.

7.32 Adventitious shoot of a Common Osier (Salix viminalis). The shoot grew out of the bark, a few days after the stem had been cut.

7.33 Adventitious shoot of a Common Osier (Salix viminalis). The shoot grew out of the cambium around the branch, a few days after the stem had been cut.

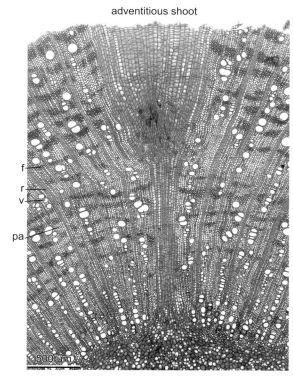

7.34 Adventitious shoot on a twig of a Fig tree (Ficus carica). The shoot developed out of a large ray. The annual rings are not recognizable.

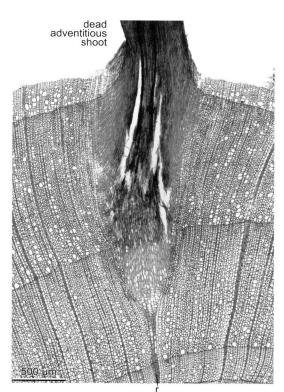

7.35 Adventitious root at a buried stem base of a Small-leaved Lime stem (Tilia cordata). The root developed out of a ray.

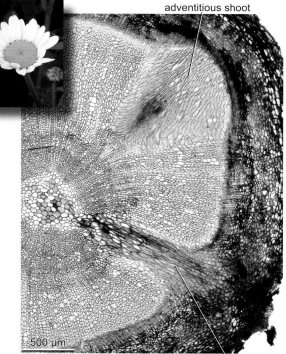

7.36 Adventitious and latent roots at the base of a hemicryptophytic herbaceous plant. Margherita (Leucanthemum vulgare; small photo: Schumacher).

7.37 Adventitious roots on a stem of Grey Alder (Alnus incana). The stem was buried twice in a riverbed and formed adventitious roots. Later, the stem was exposed by erosion.

7 MODIFICATION OF ORGANS

127

The Lateral Modification of Stems

Genetic and environmental factors modify stem cross-sections. Locally reduced or increased growth results in fluted stems. This applies to all growth forms, from trees to small herbaceous plants, and it is mainly genetically induced (7.38-7.40).

Discontinuous enlargement around the stem is mainly a result of eccentric growth. The growth discontinuity causes locally wedging rings or very eccentric stems (7.41-7.45).

Locally diverse cell differentiation creates wavy stem girths. The so-called hazel growth is well-known (7.46-7.48). A special kind of lateral modification are displaced rings in species with very large rings (7.49).

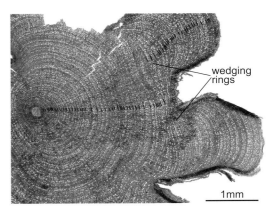

7.38 Fluted dwarf shrub stem. Growth stopped various times, resulting in stem indentations. Corema alba, Empetraceae.

7.39 Fluted and twisted Pear Tree stem (Pyrus communis).

7.40 Fluted stem of a hemicryptophytic herbaceous plant. Locally, growth reductions occurred after the second year. Mat Daisy (Raoulia tenuicaulis), Asteraceae.

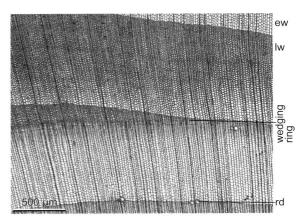

7.41 Wedging ring in a Norway Spruce stem (Picea abies). The tree grew in an avalanche track, where crown and roots were damaged. It did not have enough reserves to make a complete ring. As a result, a minimal compression wood zone and a wedging ring formed.

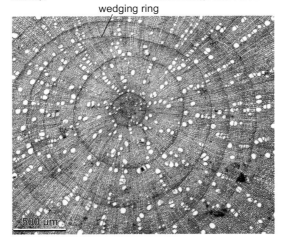

7.42 Incomplete third ring in a Hazelnut root (Corylus avellana).

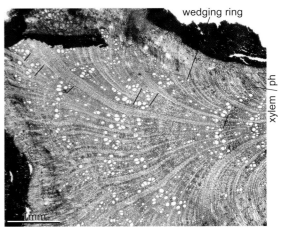

7.43 Wedging rings in a compressed dwarf shrub stem. Spiny Restharrow (Ononis spinosa).

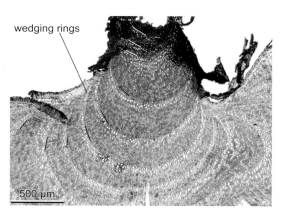

7.44 Very eccentric stem of a hemicryptophytic herb. The initial stem was partially damaged and decomposed (dark part). On one side, the remaining live part developed new conductive tissue. The rings wedge out towards the dead part. Horseshoe Vetch (Hippocrepis comosa).

7.45 Eccentric Norway Spruce branch (Picea abies). The compression wood in all rings wedges out towards the tension side (upper part).

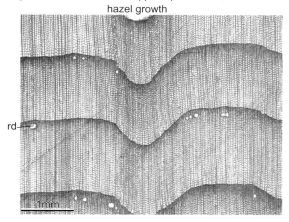

7.46 Typical hazel growth in a Norway Spruce stem (Picea abies). The rings show local indentations covering a period of several years.

7.47 Hazel growth in Common Yew (Taxus baccata). Left: Cross section. Characteristic are local indentions over many years. Right: Irregular surfaces are an expression of the indentions.

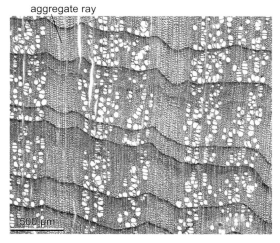
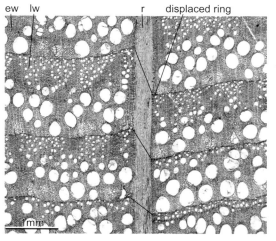

7.48 Wavy ring boundaries in a Hazelnut stem (Corylus avellana). The phenomenon is characteristic of that species.

7.49 Displaced rings along very large rays in Downy Oak (Quercus pubescens). The phenomenon is typical of oak wood.

CHAPTER 8

ANATOMICAL PLASTICITY

Size and form of a plant, its taxonomic affiliation as well as the function of its organs, creates a large variability of anatomical wood and bark structures.

Dead and living Bilberry shoots (Vaccinium myrtillus).

Wood Structural Variability in Different Families

All anatomical atlases of wood clearly demonstrate genetic differences between taxa, however, also the degree of variability within families differs greatly. Some families are very heterogeneous (8.1-8.4), others are homogeneous (8.8-8.10). Here, we demonstrate this fact using simple wood anatomical features of herbaceous plant shoots from the Ranunculaceae, tree stems from the Oleaceae, and shrub stems from the Labiatae.

There is a lot of variability in plants from the Ranunculaceae family (8.1-8.4). Taxonomic relationships are only evident in some subunits. In members of the Oleaceae family, the degree of variability is high, with respect to vessel distribution and ray composition (8.5-8.7). If one looks at wood anatomical features of Oleaceae, the taxonomic connection is not evident. There is little variability in plants from the Labiatae family (8.8-8.10). Most species are diffuse-porous and contain small, homogeneous rays. From the wood anatomical features, the taxonomic connection is, in fact, evident in Labiatae.

It would be worth including the wood anatomical features in taxonomic studies.

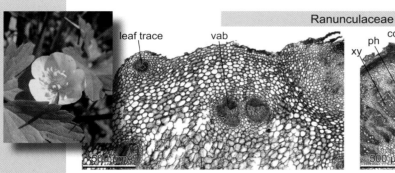

8.1 Ranunculaceae. Rhizome cross-section from the hemicryptophytic herbaceous Meadow Buttercup (Ranunculus acer) with closed vascular bundles in parenchymatous tissue.

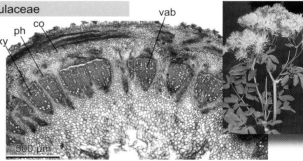

8.2 Ranunculaceae. Rhizome cross-section from the hemicryptophytic herbaceous French Meadow-rue (Thalicrum aquilegifolium) with open vascular bundles around the pith and large unlignified rays (blue). Small right: from Aeschimann et al. 2004.

8.3 Ranunculaceae. Rhizome cross-section from the hemicryptophytic herbaceous Spring anemone (Pulsatilla vernalis) with open vascular bundles with six annual growth zones and large rays with dilatations. Small left: from Aeschimann et al. 2004.

8.4 Ranunculaceae. Rhizome cross-section from the hemicryptophytic herbaceous Green Hellebore (Helleborus viridis). The stem with distinct ring boundaries and large rays between the open vascular bundles is typical of this plant. Small right: from Aeschimann et al. 2004.

Oleaceae

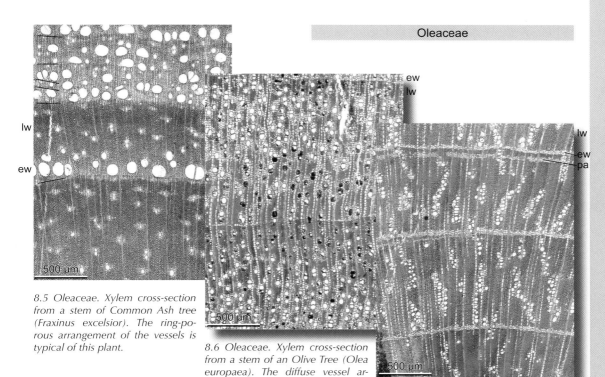

8.5 Oleaceae. Xylem cross-section from a stem of Common Ash tree (Fraxinus excelsior). The ring-porous arrangement of the vessels is typical of this plant.

8.6 Oleaceae. Xylem cross-section from a stem of an Olive Tree (Olea europaea). The diffuse vessel arrangement is characteristic.

8.7 Oleaceae. Xylem cross-section from a stem of a Pau-branco (Picconia azorica) tree. The flame-like, semi-ring-porous arrangement of the vessels is characteristic of this species.

Labiatae

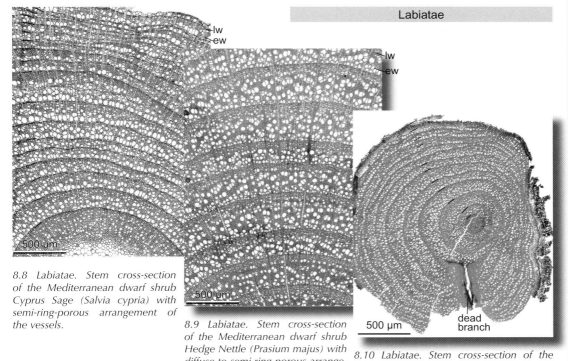

8.8 Labiatae. Stem cross-section of the Mediterranean dwarf shrub Cyprus Sage (Salvia cypria) with semi-ring-porous arrangement of the vessels.

8.9 Labiatae. Stem cross-section of the Mediterranean dwarf shrub Hedge Nettle (Prasium majus) with diffuse to semi-ring-porous arrangement of the vessels.

8.10 Labiatae. Stem cross-section of the very small, alpine dwarf shrub Wild Thyme (Thymus serpyllum) with diffuse to semi-ring-porous arrangement of the vessels.

8 ANATOMICAL PLASTICITY

Wood Structural Variability in Different Growth Forms

The wood anatomical features in different growth forms of some genera are surprisingly similar. This can be seen on tree, shrub and dwarf shrub stems. Crown size and competition between trees determine vessel size and number, as well as the number of fibers, but not the anatomical features of cell walls. This will be illustrated here on material from very different taxonomic units: the genera *Rhamnus* (8.11, 8.12), *Prunus* (8.13, 8.14), *Potentilla* (8.15, 8.16) and *Salix* (8.17-8.19).

The wood structure differs very much in heavily metamorphosed shoots, e.g. in underground rhizomes, in comparison with above-ground shoots. □

8.11 Alpine Buckthorn (Rhamnus alpina ssp. fallax). Stem wood from a 4 m tall tree with a flame-like vessel distribution.

8.12 Dwarf Buckthorn (Rhamnus pumila). Stem wood from a prostrate dwarf shrub, of the Rhamnaceae family with flame-like vessel distribution.

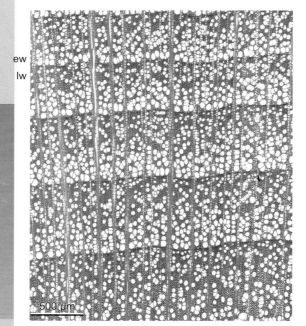

8.13 Wild Cherry (Prunus avium). Stem wood from a 15 m tall tree with a high vessel density.

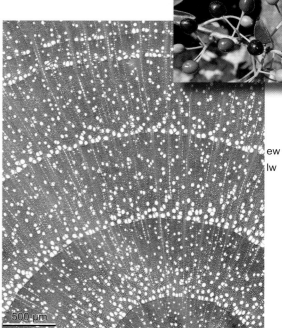

8.14 Ground Cherry (Prunus fruticosa). Stem wood from a 50 cm tall shrub with dense latewood tissue.

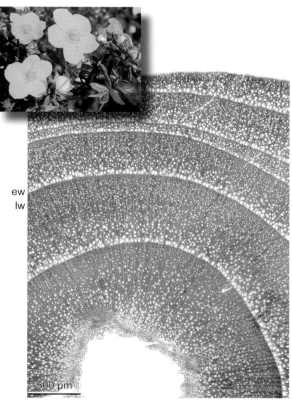

8.15 Shrubby Cinquefoil (Potentilla fruticosa). Stem wood from a 50 cm tall shrub with dense latewood tissue.

8.16 Barren Strawberry (Potentilla micrantha). Xylem from the root collar of a 5 cm tall herbaceous plant with very small rings.

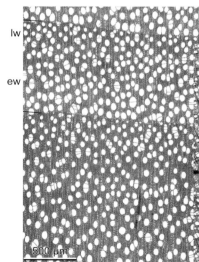

8.17 Common Osier (Salix viminalis). Stem wood from a 15 m tall tree in a wet meadow with diffuse-porous wood.

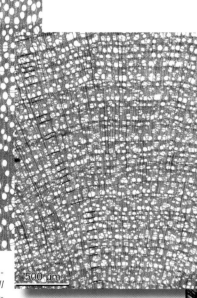

8.18 Willow (Salix breviserrata). Wood from a 50 cm tall dwarf shrub with an upright stem in a subalpine rocky scree with semi-ring-porous wood.

8.19 Dwarf Willow (Salix herbacea). Wood from a 5 cm tall dwarf shrub with a subterranean stem in an alpine snow pocket showing semi-ring porous wood. Small right: Salix herbacea in Snåsu, West Norway.

8 ANATOMICAL PLASTICITY

Wood Structural Variability Under Different Site Conditions

Site conditions modify the anatomical structure of the wood. Dendrochronological studies on conifers have clearly shown that climatic factors influence radial growth and latewood density. However, only little is known regarding the influence of climatic factors on the wood structure. It is certain that different hydrological and nutritional conditions reduce or enhance growth and modify the size of the water-conducting elements. Mechanical factors also influence growth to a large extent. Altitude is, however, of little importance.

Here, structural changes will be demonstrated that were triggered by crown size reduction (*Fraxinus*, 8.20, 8.21), lack of moisture (*Globularia*, 8.22-8.24), a reduced growing season (*Euphorbia*, 8.25-8.27) and site disturbance (*Capsella*, 8.28-8.30).

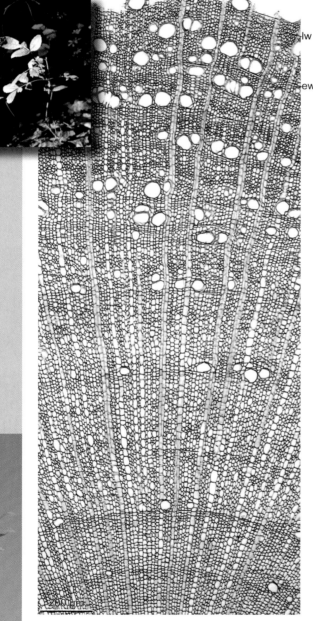

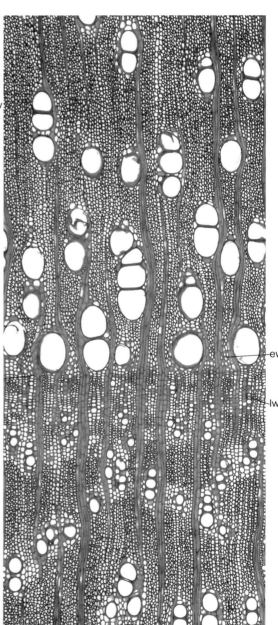

8.20 Common Ash (Fraxinus excelsior). Wood from a suppressed 20 cm tall sapling with few leaves. Characteristic are small rings, a few small vessels and thin-walled fibers. Lack of light strongly influenced the wood anatomical structures. Small left: Supressed Ash sapling. Zürich, Switzerland.

8.21 Common Ash (Fraxinus excelsior). Wood from a dominant 20 m tall tree with a big crown with wide rings, ring-porosity and dense latewood. Zürich, Switzerland.

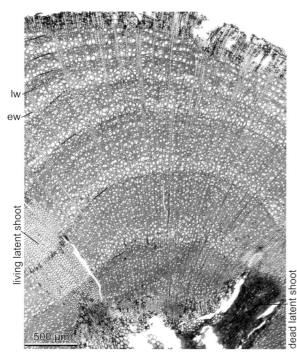

8.22 Heart-leaved Daisy (Globularia cordifolia). Wood from a dwarf garden shrub in the hilly belt showing large rings with a wide latewood zone. Optimal conditions allow optimal radial growth. Zürich, Switzerland.

8.23 Heart-leaved Daisy (Globularia cordifolia). Wood from a dwarf shrub growing in a south-facing limestone rock crevice in the hilly belt showing wavy tree-ring boundaries and narrow rings with a wide latewood zone. Frequent stem movement impedes irregular, radial growth. Walensee, Switzerland.

8.24 Heart-leaved Daisy (Globularia cordifolia). Wood from a prostrate dwarf shrub on a south-facing limestone rock in the alpine zone showing very narrow rings. A short growing season limits radial growth. Ofenpass, Switzerland. Small right: from Aeschimann et al. 2004.

continued next page

8 ANATOMICAL PLASTICITY

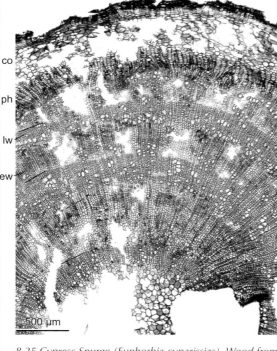

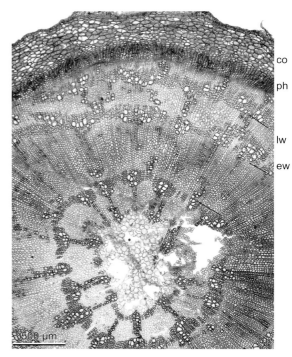

8.25 Cypress Spurge (Euphorbia cyparissias). Wood from the rhizome of a perennial plant, grown in the alpine zone. The vessels in this four-year-old shoot are surrounded by groups of fibers. Parenchyma cells form the latewood. Zermatt, Switzerland.

8.26 Cypress Spurge (Euphorbia cyparissias) Wood from the rhizome of a perennial plant, grown in a meadow of the subalpine zone. The vessels in this four-year-old shoot are surrounded by a small belt of fibers. Parenchyma cells form the latewood Zermatt, Switzerland.

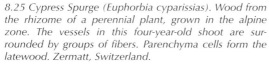

8.27 Cypress Spurge (Euphorbia cyparissias). Wood from the rhizome of a perennial plant, grown in a meadow of the montane zone. The vessels in this three-year-old shoot are surrounded by a large belt of fibers. Parenchyma cells cells form the latewood. Jura Mountains, Switzerland.

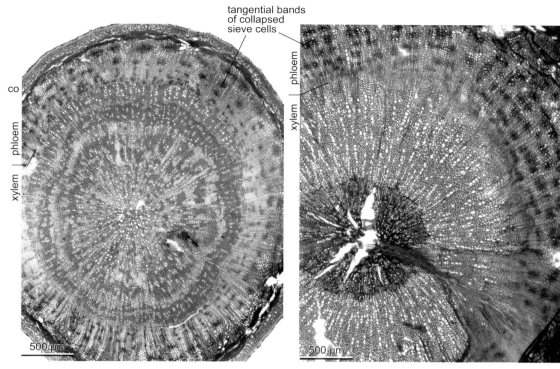

8.28 Shepherd's Purse (Capsella bursa-pastoris), a pioneer on fertile soil in the hilly belt. Wood and bark of a biannual herbaceous plant with intra-annual growth zones. The plant germinated in early fall and formed the first ring. A warm period in winter initiated radial growth by forming an earlywood zone (see small photo below). The plant was harvested in February. Aarau, Switzerland.

8.29 Shepherd's Purse (Capsella bursa-pastoris). Wood and bark of an annual herbaceous plant growing on fertile soil in the alpine zone showing regular radial growth of the xylem and tangential bands of collapsed sieve cells in the phloem. Davos, Switzerland.

8.30 Shepherd's Purse (Capsella bursa-pastoris). Wood and bark of a biannual herb showing a very small earlywood zone. Small left: The earlywood is formed only by vessels. Zürich, Switzerland. Small right: from Aeschimann et al. 2004.

8 ANATOMICAL PLASTICITY

Modification of Wood and Bark Caused by Different Shoot and Root Functions

Morphological and physiological adaptations to various shoot functions are reflected in the shoot anatomy.

In young shoots of trees, shrubs and dwarf shrubs, assimilation may have priority (8.31, 8.32, 8.34). By contrast, stabilization and water conduction appear to be more important in older plant parts (8.33, 8.35). Young shoots have a thick bark and a thin xylem. Old shoots and stems are characterized by a relatively thin bark and a relatively thick xylem.

In hemicryptophytes, the anatomy of above-ground and underground shoots is different. Annual shoots must be stable and capable of assimilation. Therefore, the stem produces much sclerenchymatous tissue and thin-walled parenchymatous tissue for chloroplasts. In contrast, the perennial rhizome must store reserves and develop a big, parenchymatous pith.

Metamorphosed shoots, for example those transformed to leaves (phylloclads) (8.36-8.39) have a very different anatomical structure. Shoots (8.40) are adapted to mechanical stress and developed a sclerenchymatic belt around the shoot. Stems of succulent plants in dry regions consist mainly of parenchyma cells (8.41).

8.31 Bilberry (Vaccinium myrtillus) with a one-year-old green, assimilating shoot, and a brown, dead shoot.

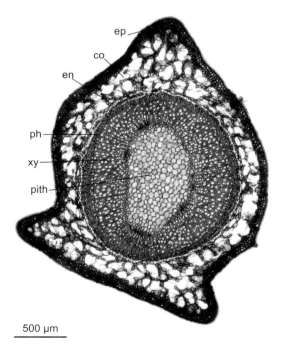

8.32 Bilberry (Vaccinium myrtillus). Cross-section of a green, assimilating one-year-old shoot. The very loose parenchymatous tissue of the cortex contains many chloroplasts. It resembles the spongy parenchyma in leaves. The xylem is very small.

8.33 Bilberry (Vaccinium myrtillus). Cross-section of an old brown stem of the same plant. The xylem expanded, and the phloem was reduced. The cortex disappeared. Water conduction and storage of carbohydrates are primary tasks of the old stem.

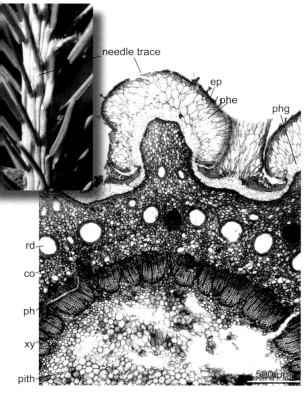
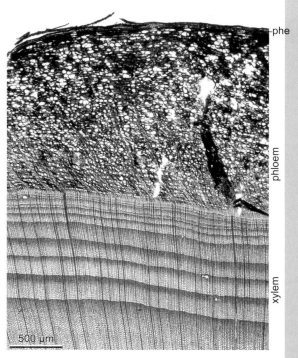

8.34 Norway Spruce (Picea abies). Cross-section of a one-year-old, vertical, green, assimilating top shoot. The cortex is much bigger than the xylem and phloem. Small left: Macroscopic view of a one year old Norway Spruce top shoot (Picea abies).

8.35 Norway Spruce (Picea abies). Cross-section of the outer part of an old, brown, suppressed stem. The xylem and phloem expanded, and the cortex disappeared.

8.36 Butcher's Broom (Ruscus aculeatus). Plant with a phylloclade with fruit and flowers. A phylloclade is a metamorphosed short shoot, the flower and fruit is formed on a side branch (see also p. 202).

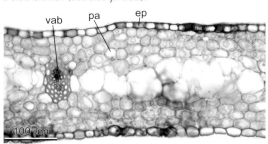

8.37 Butcher's Broom (Ruscus aculeatus). Cross-section of a phylloclade. The leaf-like structure is typical; palisade cells are missing. Most cells originally contained chloroplasts. The dense sclerenchymatous tissue around the vascular bundles stabilizes the leaf-like structure.

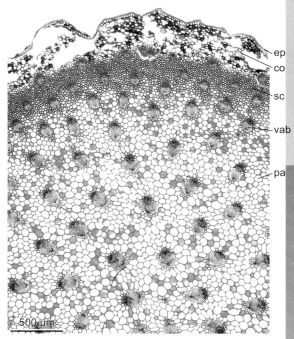

8.38 Butcher's Broom (Ruscus aculeatus). Stem cross-section. The irregular distribution of closed, vascular bundles in the parenchymatous tissue, the dense sclerenchymatous belt and the surrounding assimilating, thin-walled external tissue are characteristic of this plant.

continued next page

8 ANATOMICAL PLASTICITY

141

8 Anatomical Plasticity

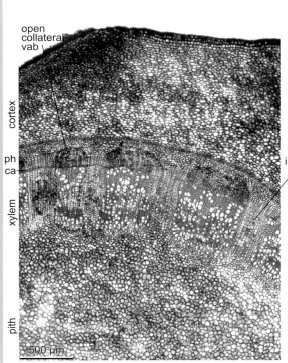

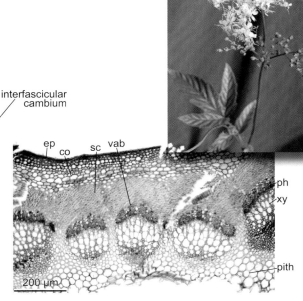

8.39 Meadowsweet (*Filipendula ulmaria*). Cross-section of a perennial rhizome. The large, parenchymatous tissue, the big pith inside and the cortex outside of the xylem/phloem ring are typical of rhizomes.

8.40 Meadowsweet (*Filipendula ulmaria*). Cross-section of an annual shoot with vascular bundles and a very dense sclerenchymatous belt in response to mechanical stress.

8.41 Prickly Pear (*Opuntia sp.*). Cross-section of a fleshy lobe showing small, radial fiber/vessel strips, which consist of small cells between extended parenchymatous tissue with big cells. This species grows at high altitudes of the Andes (Altiplano) and forms annual rings. Small right: Opuntia in the Argentinean Altiplano, about 4000 m a.s.l.

Chapter 9

Wood Structural Modifications Caused by Weather and Climate

The distribution and size of vessels, fibers and parenchyma are a reaction to abiotic site conditions.

Wind-stressed Larch trees (Larix decidua). Bernina pass, Switzerland.

Major Wood Anatomical Types in Different Climatic Regions

Every biome on the globe contains characteristic growth forms, for example, large trees with big leaves in the tropics and little shrubs with small leaves in the desert. In most angiosperms, the anatomical structure reflects morphological adaptations (Baas and Schweingruber 1987). Such adaptations do not exist in conifers, in angiosperms with successive cambia and in climbing plants. In the following, cross-sections of angiosperm wood are presented.

The wood of many **tropical rain forest** (9.1) trees contains big vessels that are accompanied by groups of parenchyma cells (9.2-9.4). However, intensive water transport is also possible in conifer-like tissues (9.5).

9.1 Tropical rain forest in Costa Rica.

9.2 Flat-top Acacia tree (Acacia abyssinica), Fabaceae from the tropical region of East Africa. Characteristic are the big vessels (>100 µm) that are surrounded by paratracheal parenchyma.

9.3 Zebrawood tree (Microberlinia sp.), Fabaceae from tropical West Africa, with 5200 mm annual rainfall. The big vessels (>200 µm) that are surrounded by paratracheal parenchyma are characteristic of this tree. The dark zone is only a layer of phenolic substances, not a tree-ring boundary photo: Worbes).

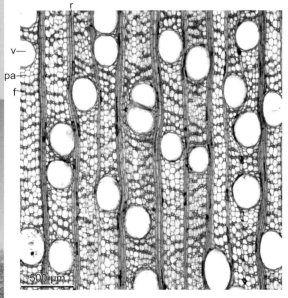
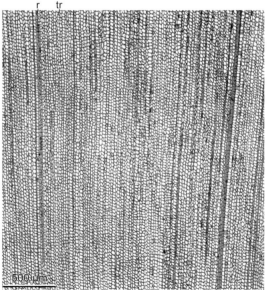

9.4 Pseudobombax mungabae tree, Bombaceae from tropical Brazil, with four to six arid months per year. The big vessels (approx. 100 µm) are typical of this tree; (photo:Worbes).

9.5 Drymis piperita tree, Winteraceae from tropical New Guinea. The absence of vessels is typical for the Winteraceae family.

On dry sites in **arid and semi-arid** locations (9.6), tree and shrub wood mostly has small vessels and apotracheal and paratracheal parenchyma (9.7 - 9.10)

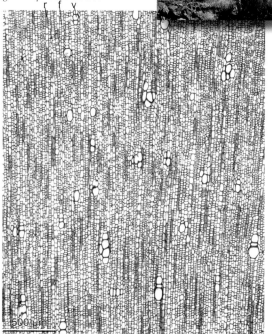

9.6 Arid tree-shrub mountain vegetation (Pinyon) in the White Mountains of the Sierra Nevada. In the foreground is a green Ephedra shrub.

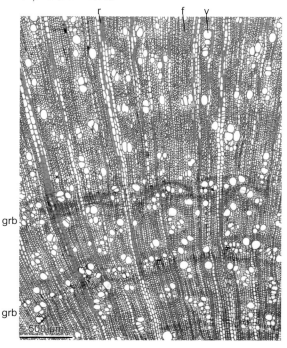

9.7 Withania adpressa shrub, Solanaceae, from the arid climate of Morocco, showing medium-sized vessels (<50 μm).

9.8 Euphorbia pescatoria shrub, Euphorbiacea, from the arid climate of Morocco, showing medium-sized vessels (>50 μm) are embedded in very thin-walled fiber tissue and parenchyma.

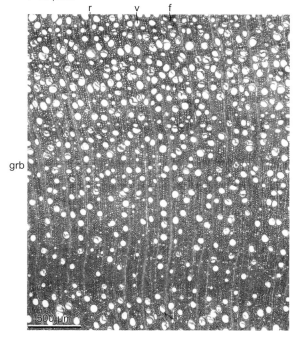

9.9 Guijera parviflora shrub, Rutaceae, from the arid climate of the Nullarbour Plain of Australia, showing medium-sized vessels (approx. 50 μm) embedded in thick-walled fiber tissue.

9.10 Small Nolana sp. shrub, Nolanaceae, from the Athacama Desert, Chile, showing medium-sized vessels (>50 μm) embedded in thick-walled fiber tissue.

continued next page

9 Modifications Caused By Weather and Climate

9 Modifications Caused By Weather and Climate

Tree and shrub wood from **temperate regions** with a seasonal climate (9.11) has medium-sized vessels and apotracheal parenchyma (9.12-9.15).

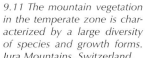

9.11 The mountain vegetation in the temperate zone is characterized by a large diversity of species and growth forms. Jura Mountains, Switzerland.

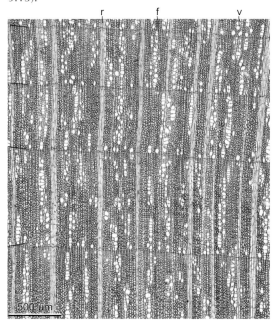

9.12 Small Common Holly tree (Ilex aquifolium), Aquifoliaceae, from the temperate climate of Switzerland, showing medium-sized vessels (30-50 μm) arranged in radial multipels.

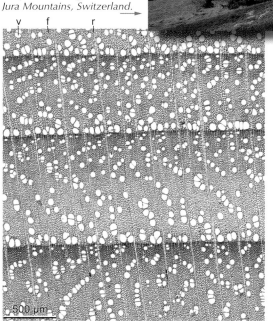

9.13 Wild Cherry tree (Prunus avium), Rosaceae, from the temperate climate of Switzerland, showing semi-ring-porous wood with medium-sized vessels (30-60 μm).

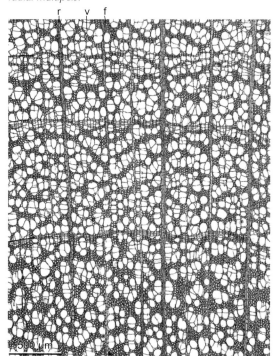

9.14 Small Red-berried Elder tree (Sambucus racemosa), Rosaceae, showing numerous medium-sized vessels arranged in groups (30-60 μm). Swiss temperate climate.

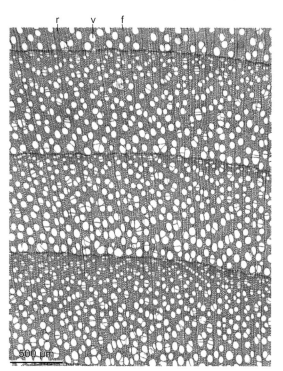

9.15 Small Purple Osier tree (Salix purpurea), Salicaceae, from the temperate climate of Switzerland, showing numerous, medium-sized solitary vessels (30-60 μm).

Shrub wood from **boreal and subarctic zones** (9.16) has numerous, very small vessels and apotracheal parenchyma (9.17-9.20).

9.16 Dwarf shrub vegetation in the Canadian arctic. Prostrate willows grow mainly in the moist channels between ice polygons.

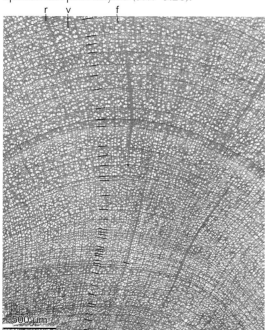

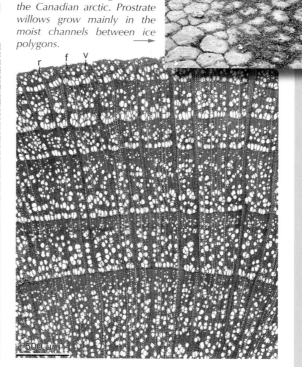

9.17 Small Rhododendron parviflora shrub, Ericaceae, from Eastern Siberia, growing at the Northern timberline of a boreal climate, showing numerous, very small vessels (<30 μm).

9.18 Small Steven's Meadowsweet (Spiraea stevenii) shrub, Rosaceae, from Eastern Siberia, Magadan, at the Northern timberline, showing semi-ring-porous wood with numerous small vessels (approx. 50 μm).

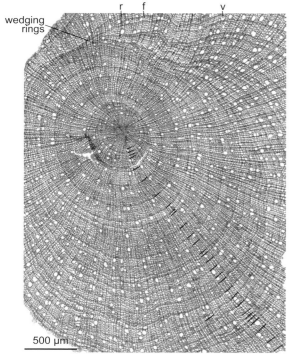

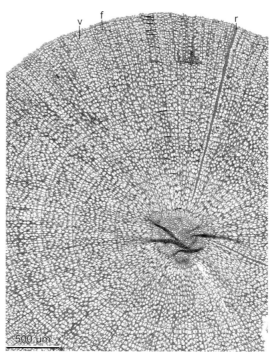

9.19 Dwarf Birch shrub (Betula nana), Betulaceae, from Western Siberia, growing at the Northern timberline of a boreal climate, showing very small vessels (<50 μm).

9.20 Leather-leaf (Chamaedaphne calyculata), Ericaceae, dwarf shrub, from Western Siberia, growing in a bog, showing numerous, very small vessels (<30 μm).

9 MODIFICATIONS CAUSED BY WEATHER AND CLIMATE

Modification of the Annual Tree-Ring Formation Caused by Seasonal Climatic Changes

The formation of annual tree-ring boundaries is closely related to seasonal, climatic changes. In the humid tropics, with little seasonal temperature variation, the presence of tree-ring boundaries is limited. Worbes (1994) found that about 50% of all trees in the Amazon Basin do not have annual tree-ring boundaries (9.21, 9.22). By contrast, most species growing in seasonal climates of temperate (9.25, 9.26) and boreal zones (9.27, 9.28) show distinct annual ring boundaries. In deserts, ring boundary formation is related to rainfall (9.23, 9.24). It seldom occurs in deserts with a high winter temperature, for example, in the Sahara (>10 °C); yet, tree-ring boundaries are distinct in deserts with a continental climate and cold winters, such as in Central Asia. Sporadic rainfall creates "rain rings" rather than "annual growth rings" (Schweingruber and Poschlod 2005).

It is important to know that the clarity of rings varies greatly in all climates. Intra-annual climatic variations affect the capacity of the tree to form rings. In the following a few characteristic tree-ring boundaries from tropical, arid, temperate and boreal zones are shown.

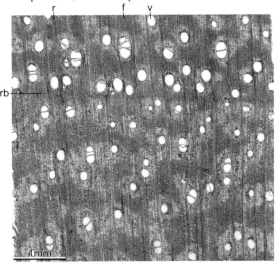

9.21 Indistinct tree-ring structures in a tropical tree. The boundary can be identified by a slight difference in vessel size. Flat-top Acacia (Acacia abyssinica).

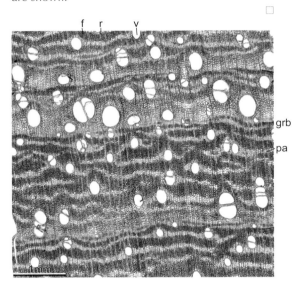

9.22 Distinct tree-ring boundary in a tropical tree. The boundary is indicated by the difference in latewood and earlywood densities. African Teak (Pterocarpus angolensis).

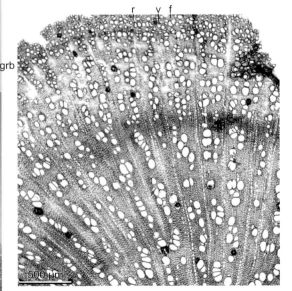

9.23 Indistinct tree-ring boundary in a dwarf shrub from the Sahara desert. The boundary is partially indicated by a fiber band with thick cell walls. Malcolmia aegyptica, Morocco.

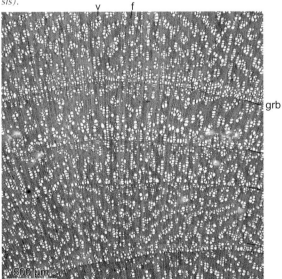

9.24 Distinct tree-ring boundaries in a dwarf shrub from the Athacama desert (Chile). The boundaries are identifiable mainly by slight semi-ring-porosity. Schinus piliferus.

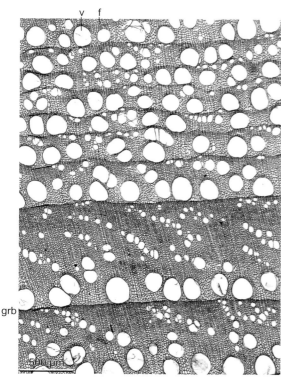

9.25 Distinct annual tree-ring boundaries in a tree from the temperate region of the Alps. The boundary is indicated by ring-porosity. Sweet Chestnut (Castanea sativa). Ticino, Switzerland.

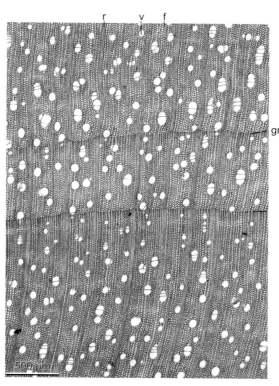

9.26 Distinct annual tree-ring boundaries in a tree from the temperate region of the Alps. The boundary is indicated by a small fiber band with thick cell walls in the latewood. Sycamore (Acer pseudoplatanus). Ticino, Switzerland.

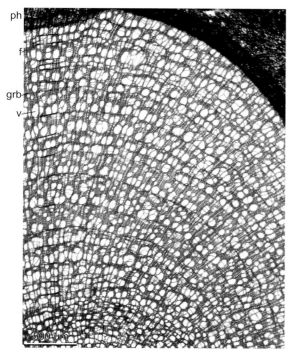

9.27 Distinct annual tree-ring boundaries in a dwarf shrub from the alpine zone in the Alps. The boundaries are marked by slight semi-ring-porosity. Dwarf Willow (Salix herbacea). Davos, Switzerland.

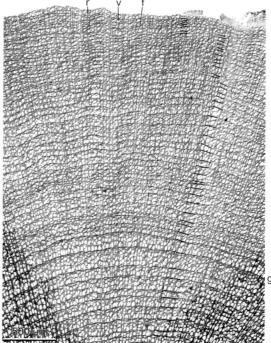

9.28 Distinct annual tree-ring boundaries in a dwarf shrub from the alpine zone in the Polar Urals. The boundaries are marked by slight semi-ring-porosity and a few fibers with thick latewood cell walls. Mountain Azalea (Loiseleuria procumbens). Davos, Switzerland.

9 Modifications Caused By Weather and Climate

Modification of the Annual Tree-Ring Formation Caused by Seasonal Climatic Changes: The Genetic Component

The production of tree rings is not only environmentally but also genetically determined. This applies to all climates. In temperate climates, in regions with neither very cold winters nor very hot summers, such as Southern Europe or the Californian coast, many species have well-defined tree-ring boundaries, but many others only have indistinct growth zones.

The variable genetic predisposition is demonstrated here for trees in a Southern Portuguese plantation at Monchique, 900 m above sea level. The mean average January and July temperatures there are 9 °C and 23 °C, respectively. Some species have distinct (9.29, 9.32, 9.34), some indistinct (9.30, 9.31) and others missing ring boundaries (9.33).

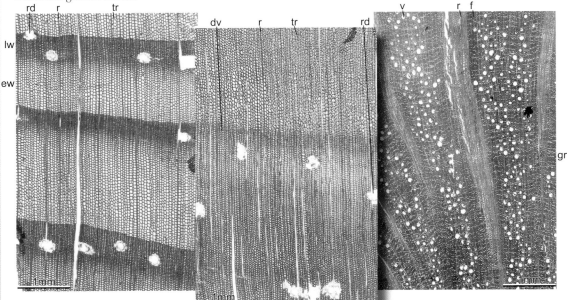

9.29 Maritime Pine (Pinus pinaster), Pinaceae. Distinct tree-ring boundaries. The species is endemic to Mediterranean.

9.30 Umbrella Pine (Pinus pinea), Pinaceae. Indistinct ring boundaries. Endemic to Mediterranean.

9.31 Kermes Oak (Quercus coccifera), Fagaceae. Indistinct ring boundaries. Endemic to Mediterranean.

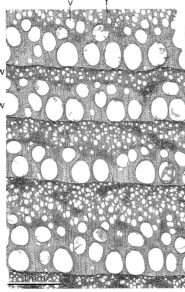 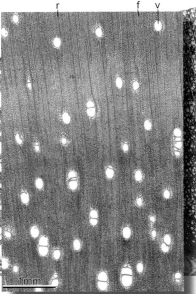 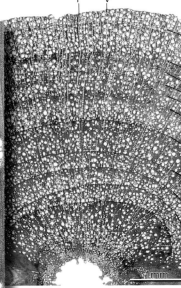

9.32 Sweet Chestnut (Castanea sativa), Fagaceae. Distinct ring boundaries. Endemic in Eastern Europe.

9.33 Eucalyptus sp., Myrtaceae. In this case no distinct ring boundaries.

9.34 Heather (Calluna vulgaris). Ericaceae. Distinct ring boundaries. Dwarf shrub endemic to Europe.

Generally, annual tree rings in non-tropical woodlands are divided into earlywood and latewood. In most types of wood the vessel area is greater in the earlywood than in the latewood (9.35-9.38). However, there are also many species which do not show clear differences in vessel area between early and latewood (9.39, 9.40). Increased water conduction at the beginning and little water transport at the end of the growing season can trigger vessel size differences but some species developed other mechanisms to regulate variable water availability. Here, the genetic component of earlywood and latewood vessel size and distribution is demonstrated.

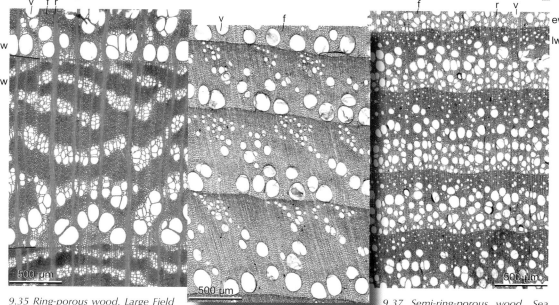

9.35 Ring-porous wood. Large Field Elm shrub (Ulmus minor). The earlywood is characterized by large vessels, and tangential groups of small vessels in the latewood.

9.36 Ring-porous wood. Sweet Chestnut tree (Castanea sativa). The earlywood is characterized by large vessels and flame-like groups of vessels in the latewood.

9.37 Semi-ring-porous wood. Sea Buckthorn shrub (Hippophae rhamnoides). The earlywood vessels are bigger than those in the latewood. The optimal vessel size is in the middle of the ring.

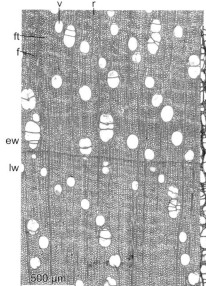

9.38 Diffuse-porous wood. Common Walnut tree (Juglans regia).

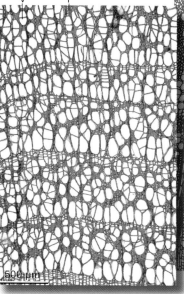

9.39 Semi-ring-porous wood. Stunted Willow (Salix retusa). Prostrate, alpine dwarf shrub. The earlywood vessels are bigger than those in the latewood.

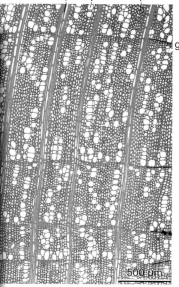

9.40 Diffuse-porous wood. Common Holly shrub (Ilex aquifolium).

9 Modifications Caused By Weather And Climate

Modification of the Xylem due to Intra-Seasonal Variations: Ecological, Climatic and Individual Compontents

The genetic identity of each taxon is extremely flexible. The interannual distribution, size and number of vessels, fibers and parenchyma, as well as the cell wall thickness of vessels and fibers, are all very variable. This flexibility permits plants to react to different site conditions and short-term environmental changes. For conifers, the changes in ring width and latewood density, the phenomena of cell collapse and callus are well known and understood. By contrast, structural variations can hardly be explained.

Here, a few examples of structural changes in angiosperms are shown, such as differences in vessel size and the frequency of parenchyma bands: *Laburnum* growing under different site conditions (9.41, 9.42), differences in the vessel size of *Castanea* and *Salix* within a tree-ring series (9.43, 9.44), and variable tangential bands of parenchyma cells in a *Quercus* specimen (9.45, 9.46).

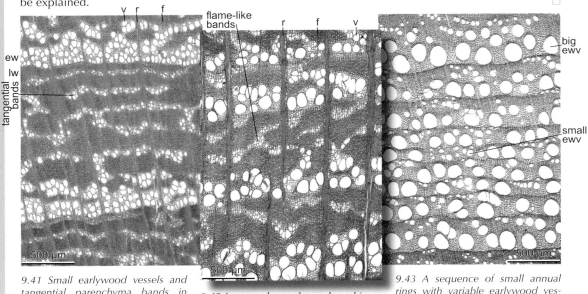

9.41 Small earlywood vessels and tangential parenchyma bands in the latewood of a Golden Rain tree (Laburnum anagyroides), grown in a garden with deep soil. Ticino, Switzerland.

9.42 Large earlywood vessels and irregular parenchyma bands in the latewood of a Golden Rain tree (Laburnum anagyroides), grown at a dry site. Ticino, Switzerland.

9.43 A sequence of small annual rings with variable earlywood vessel size. Sweet Chestnut (Castanea sativa). Ticino, Switzerland.

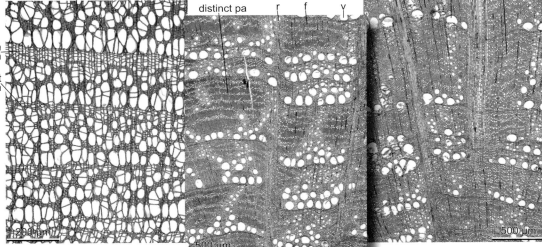

9.44 Annual rings with variable vessel size in the earlywood and different numbers of parenchyma cells in the latewood of an alpine, prostrate dwarf shrub. Stunted Willow (Salix retusa). Zermatt, Switzerland.

9.45 Annual rings with distinct tangential parenchyma bands in latewood on the tension side. Downy oak (Quercus pubescens). Valais, Switzerland.

9.46 Annual rings with indistinct tangential bands of parenchyma cells in the latewood on the compression side of the same plant. Downy oak (Quercus pubescens). Ticino, Switzerland.

Chapter 10

Wood Structural Modifications Caused by Extreme Events

Extreme weather events, fire, mechanical damages, parasites and lack of light and water lead to the formation of special anatomical features.

Tamarack (*Larix laricina*) stem with overgrown cracks and frost ribs.

Lack of Light

Dense shading is defined as a light intensity of less than 2% of the available sunlight (Larcher 2001). This lack of light inhibits growth in every sense: shoot elongation is reduced; young plants remain small, leaf mass is low and ramification inhibited; the crowns of old pines are reduced in size due to shading from neighbouring trees.

Within the xylem, lack of light affects many anatomical features: cell division is reduced, and the differentiation of cells and cell walls is affected.

Here, the xylem of normally grown trees, and of ecological dwarfs in deep shadow, will be compared.

We show dominant and suppressed conifers (10.1, 10.2), ring porous and diffuse porous angiosperms (10.3-10.6), and conifers and angiosperms under light stress (10.7-10.11). A sudden increase in light, for example after thinning operations or in clearings after a wind throw, is also reflected in the xylem (10.12, 10.13). In fact, all growth processes return to normal.

10.1 Large rings are characteristic of trees grown under normal light conditions. The earlywood is wide and light, and the latewood narrow and dense. Norway Spruce (Picea abies).

10.2 Small rings are characteristic of trees grown under reduced light conditions. The earlywood is very narrow, and the latewood is relatively wide and dense. Norway Spruce (Picea abies). Small right: Supressed spruce in a beech forest.

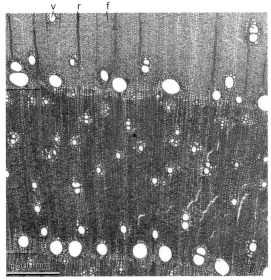

10.3 Rings of a 15 m tall Ash tree (Fraxinus excelsior), grown under normal light conditions, showing wide rings with large earlywood vessels, and wide, dense latewood.

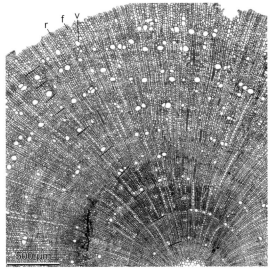

10.4 Narrow rings of a 40 cm tall Ash (Fraxinus excelsior), grown under a dense canopy of trees, showing only few small vessels embedded into thin-walled fiber/parenchyma tissue.

10.5 Wide rings of a 15 m tall, Wild Cherry tree (Prunus avium), grown under normal light conditions, showing wide, semi-ring-porous rings.

10.6 Narrow rings of a 40 cm tall, Wild Cherry sapling (Prunus avium), grown under a dense tree canopy, showing a few small vessels, indistinct ring boundaries and a great number of large rays.

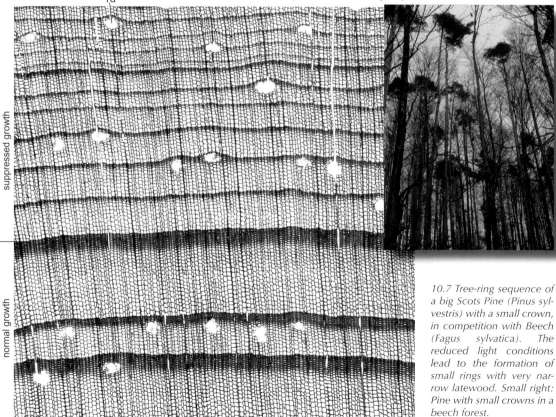
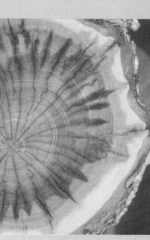

10.7 Tree-ring sequence of a big Scots Pine (Pinus sylvestris) with a small crown, in competition with Beech (Fagus sylvatica). The reduced light conditions lead to the formation of small rings with very narrow latewood. Small right: Pine with small crowns in a beech forest.

continued next page

10 MODIFICATIONS CAUSED BY EXTREME EVENTS

10 Modifications Caused by Extreme Events

Left: 10.8 Beech cohort (Fagus sylvatica) in a coppice stand. After resprouting, many stems died due to density dependent self-thinning. The period prior to death is characterized by leafless, live stems and an abnormal wood anatomy.

Right: 10.10 Dead spruce branches in a plantation.

10.9 Tree-ring sequence from a severely suppressed, almost leafless Beech tree (Fagus sylvatica) in a dense beech stand. The last rings contain very few vessels and are barely separated.

10.11 Tree-ring sequence of a dead Stone Pine branch (Pinus cembra), grown under the dense canopy. Marginal growing conditions are reflected by narrow rings and missing latewood.

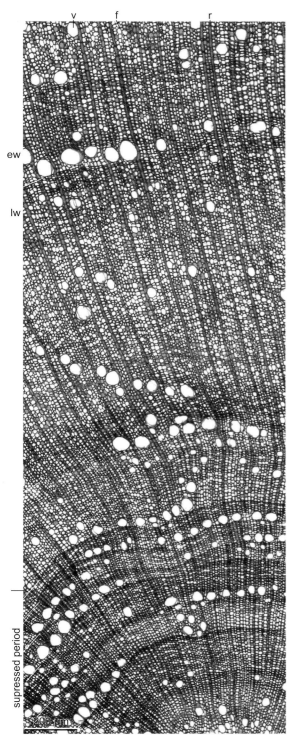
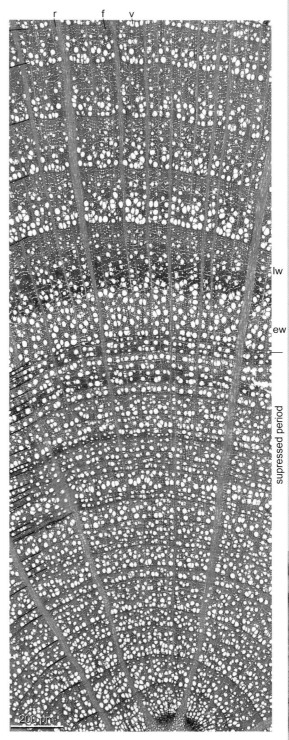
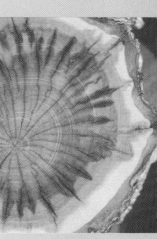

10.12 Tree-ring sequence of a Sweet Chestnut sapling (Castanea sativa), which started life in the deep shadow of a dense canopy. Following the death of a competing large tree, the sapling immediately produced much bigger rings.

10.13 Tree-ring sequence of a Beech branch (Fagus sylvatica), formerly grown under a dense canopy. After the neighbouring tree had been felled, the old branch immediately increased its growth.

Severe Frost

At the upper timberline, frost damage to leaves and needles is frequent (Körner 1999; 10.14). Here, only the phenomena of xylem frost rings will be discussed. Either the formation of ice crystals or high tension in the xylem near the cambial zone induce frost rings (Schweingruber 2001), which are characterized by a small zone of collapsed cells, a rather large zone of callus cells and bent rays.

The presence of frost rings enables the dendrochronological reconstruction of frost events. Located at the beginning of a ring, they indicate an extremely cold period prior to the growing season. Such a phenomenon produces a real frost ring. In young trees, at the timberline, frost rings in the earlywood and latewood also form during the growing season, e.g. on the first warm day following an extremely cold period, and are induced by high tension in the xylem. In effect, the thawed plant top begins to transpire and requires water which, however, cannot be provided by the still frozen root system and lower stem, the phenomenon being called "Frost-drought". Also extreme drought causing intensive water loss in the cambial area, e.g. after root exposure and defoliation, may induce "frost-type" rings.

Here, we present frost rings in a young conifer (10.15) and two broad leaf trees (10.16, 10.17). Frost ribs are a result of delayed wound healing. Periodic, alternate freezing and thawing prevents the wound margins from growing together and closing (10.18-10.20). Cracks along rays are probably induced by extreme xylem tension. As long as the ray cells are alive they fill the crack with callus tissue (10.21).

Left: 10.14 Lack of snow cover caused frost damage on Red Bearberry leaves (Arctostaphylos uva-ursi).

10.15 Repeated frost rings indicated by callus cells (cal) in a young Siberian Larch stem (Larix sibirica) in Mongolia. The young, thin stems are very susceptible to sudden temperature changes.

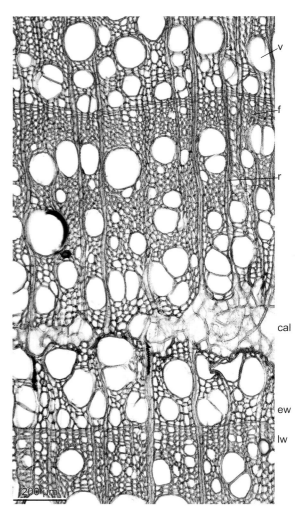

10.16 Frost ring (cal) in a semi-porous angiosperm. The frost occurred immediately after the earlywood vessels had formed. Sea Buckthorn (Hippophae rhamnoides) near the timberline in the Alps.

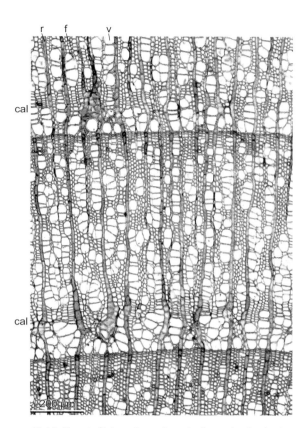

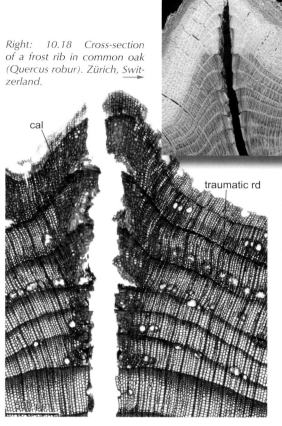

Right: 10.18 Cross-section of a frost rib in common oak (Quercus robur). Zürich, Switzerland.

10.17 Two indistinct frost rings (cal) at the beginning of the earlywood in diffuse-porous Nothofagus pumilio wood in Patagonia, Andes.

10.19 Cross-section of a frost rib in a small Siberian Larch stem (Larix sibirica). The radial crack did not close very well. It split along the callus tissue after shrinking. Eastern Siberia.

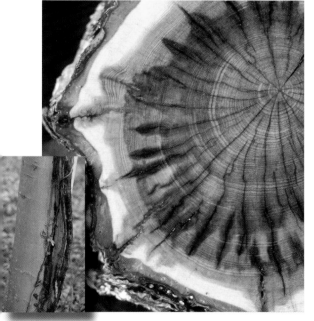

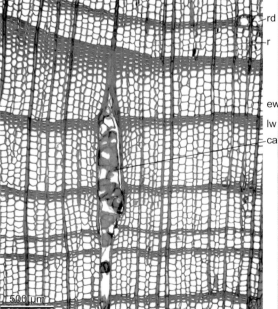

10.20 Cross-section of a Tamarack stem (Larix laricina) with many radial cracks and overgrown frost ribs. On the northern timberline in Quebec, Canada. Small above: Frost rib on a Sycamore stem (Acer pseudoplatanus). Zürich, Switzerland.

10.21 Radial crack along a ray in a Siberian Larch (Larix sibirica). The live ray cells produced callus cells (cal) and filled the crack.

10 Modifications Caused by Extreme Events

Drought and Drainage

Periodic drought during the growing season cause density fluctuations of the wood. They are very frequent in trees growing in climates with little summer rainfall, as in monsoon and Mediterranean climates (10.22, 10.23).

Plants on very-well-drained sites are usually exposed to drought. Hence, in riverbeds, after dry periods, ash and alder trees are subject to cell collapses (10.24-10.26). Very high transpiration tension would seem to induce radial cracks in the earlywood of fast-growing conifers (Cherubini et al. 1997). When irrigation systems in dry agricultural regions are turned off, the trees suffer from drought. Both start and intensity of a period of water shortage are reflected in wood anatomical structures (10.27). By contrast, trees in bogs (10.28) profit from drainage; a lower water table permits the formation of a more extensive root system, resulting in wider rings (10.29).

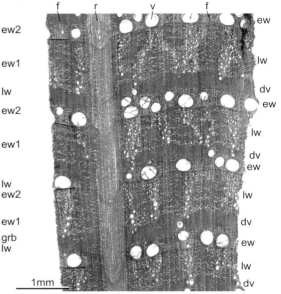

10.22 Periodic intra-annual density variations in a Smooth-leaved Pine (Pinus leiophylla), grown in the monsoon climate. An early summer drought retards radial growth and triggers cell wall thickening. Summer rains speed up growth and produce a second earlywood before terminal growth starts in the autumn. Arizona, USA.

10.23 Periodic intra-annual density variations in a Downy Oak (Quercus pubescens), grown in the Mediterranean climate. An early summer drought triggers the formation of a dense tangential fiber band. Genova, Italy.

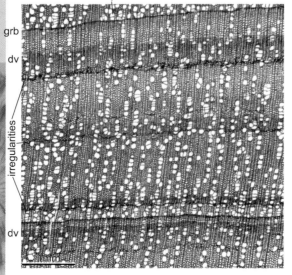

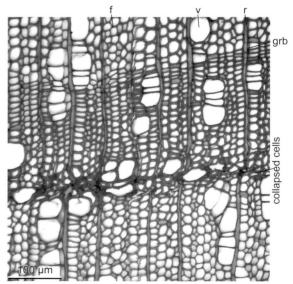

10.24 Intra-annual growth irregularities in a Grey Alder stem (Alnus incana), grown in a periodically dry riverbed. Maggia, Switzerland.

10.25 Intra-annual growth irregularities, consisting of fibers with variable cell wall thickness, collapsed fibers and vessels. Maggia, Switzerland.

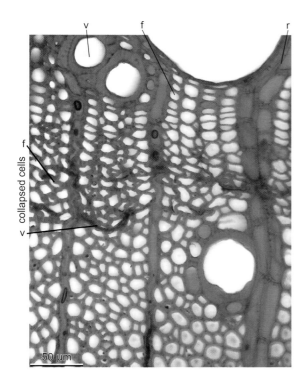
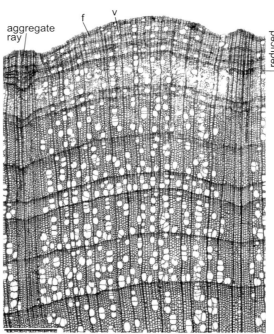

10.26 Intra-annual fiber and vessel collapse in Ash trees (Fraxinus excelsior). This was caused by an extremely low river water level in the dry summer of 1976. Ticino, Switzerland.

10.27 Terminal rings in a dead Grey Alder (Alnus incana), standing along an abandoned irrigation channel. The last rings reflect an extreme water shortage. Valais, Switzerland.

10.28 Extremely slow-growing Norway Spruce (Picea abies) in a bog. Northern Prealps, Switzerland.

10.29 Radial growth increased immediately after bog drainage. A lower water table allows oxygen to penetrate to greater depth as prerequisite for root growth. Norway Spruce (Picea abies). Northern Prealps, Switzerland.

10 Modifications Caused by Extreme Events

Defoliation by Insects

Numerous insects defoliate plants. A reduced photosynthetically active leaf area results in anatomical changes in the xylem. Here, the effects of three examples of defoliation are described.

Every seven to ten years, at the beginning of the growing season, **Larch Budmoth** (*Zeiraphera diniana*) outbreaks heavily defoliate European Larches (*Larix decidua*) in the Alps (Baltensweiler and Rubli 1999; 10.30, 10.31). Anomalous rings are the result of missing photosynthesis: in the first year of the insect attack, the earlywood consists of a few thin-walled tracheids and very few very thin-walled latewood cells. Over the following two to four years, wood production is reduced but most trees survive (10.32, 10.33).

Small below: 10.30 Destroyed leaves on short shoots of European Larch (Larix decidua; photo: Maksymov).

10.31 European Larches (*Larix decidua*) that were defoliated by the Larch Budmoth (*Zeiraphera diniana*) in early July. The accompanying stone pines are not affected. Engadin, Switzerland.

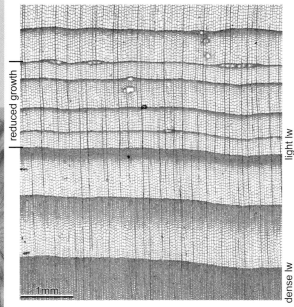

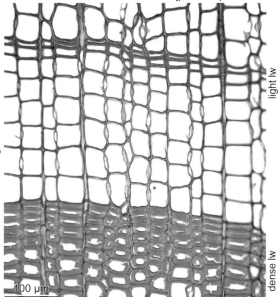

10.32 Sudden growth reduction after a Larch Budmoth infestation. In this case, the regeneration period took five years. European Larch (*Larix decidua*), Engadin, Switzerland.

10.33 Tree rings before and after a Larch Budmoth infestation. The ring of the year when the attack started, has latewood consisting of thin-walled tracheids. European Larch (*Larix decidua*). Engadin, Switzerland.

In the Rocky Mountains, over a period of several years, **Spruce Budworm** (*Choristoneura* sp.) periodically defoliates Douglas. Fir (*Pseudotsuga menziesii*) and Engelmann Spruce trees (*Picea engelmannii*) (Swetnam and Lynch 1993). Many trees die, and those that survive show typical stress reactions in the wood anatomy. Characteristic are small rings with a reduced latewood zone and tangential rows of resin ducts (10.34-10.36).

Right: 10.34 Selective Spruce Budworm (Choristoneura fumiferana) infestation of Douglas Fir (Pseudotsuga menziesii). Netherland, Colorado, USA.

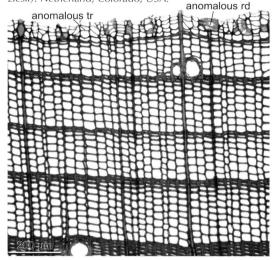

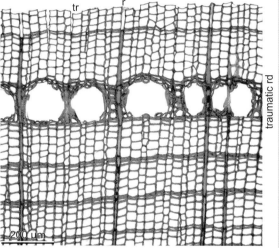

10.35 Anomalous ring after a Spruce Budworm attack, prior to death. Douglas Fir (Pseudotsuga menziesii). Netherland, Colorado, USA.

10.36 Rings after a Spruce Budworm attack, prior to death, with a tangential row of resin ducts. Netherland, Colorado, USA.

In Central Europe, the **Cockchafer** (*Melolontha melolontha*, 10.37) periodically defoliated oaks (*Quercus petraea* and *Q. robur*) just after bud break and earlywood formation (Christensen 1987). The defoliation is reflected by a small, tangential, flat row of fibers. The formation of a second generation of leaves corresponds to a narrow, false, diffuse-porous ring (10.38). □

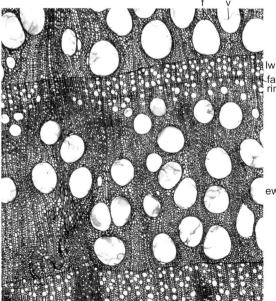

10.37 Male Cockchafer (Melolontha melolontha) on an oak (photo: Wermelinger).

10.38 Defoliation by the Cockchafer leads to the production of a false ring in a Common Oak (Quercus robur). Ticino, Switzerland.

Defoliation Caused by Chemical Pollution and Nuclear Radiation

Natural and man-made gaseous emissions cause leaf damage, defoliation (10.39, 10.40) and wood decay (10.41). The cell differentiation in the cambial zone remains normal. In conifers and angiosperms, a reduction in the leaf area, for example due to sulphur dioxide, is followed by reduced radial growth. Dead wood is delignified. Annual conifer rings are smaller and often contain less latewood than during unpolluted periods (10.42). Birches react differently to sulphur dioxide: their latewood seems to be unaffected, but the earlywood zone is reduced (10.43).

Radioactive radiation also affects cell differentiation. This has been shown for trees in the immediate neighbourhood of the exploded nuclear power station Chernobyl in the Ukraine. Disturbed mitotic cell division in the cambial zone caused irregularities in the radial orientation of the earlywood (10.44). Reduced growth results from leaf loss (Musaev 1996; 10.45).

10.39 Natural fumigation of Engelmann Spruce trees (Picea engelmannii) with sulphuric acid near hot springs in Yellowstone Park, USA. The trees around the newly formed pond died.

10.40 Selective killing of trees by sulphuric acid 140 km south of the smelter Norilsk in Siberia. One Siberian Spruce (Picea obovata) is unaffected, another one is half dead. Most silver birches (Betula pendula) are almost dead. All Siberian larches (Larix sibirica, background) are dead.

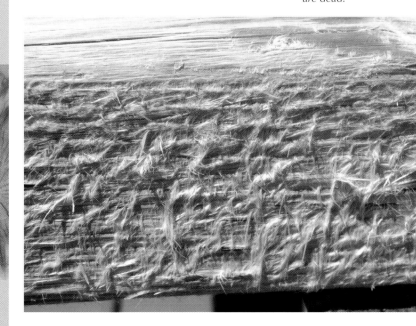

10.41 Cellulose fibers on a board near a hot spring. The sulphuric acid vapours delignify the dead wood. It is a „natural cellulose factory". Yellowstone National Park, USA.

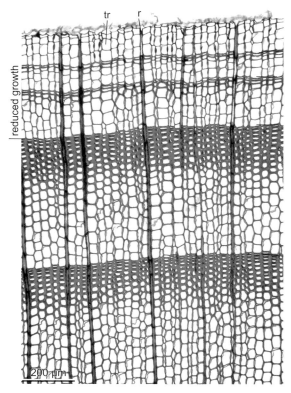

10.42 Growth reduction in a Siberian Larch tree (Larix sibirica) that was severely affected by sulphuric acid, near the smelter Norilsk, Siberia.

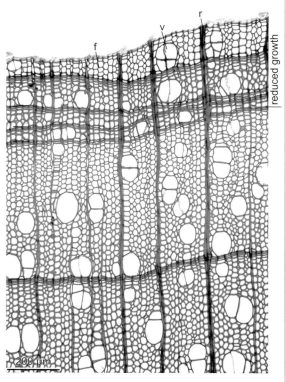

10.43 Reduced earlywood in Silver Birches (Betula pendula) that were severely affected by sulphuric acid, near the smelter Norilsk, Siberia.

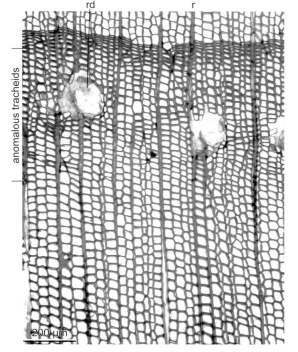

10.44 Radial tracheid rows in Scots Pine (Pinus sylvestris) were distorted by intensive gamma radiation in May 1986, near the exploded nuclear power station Chernobyl.

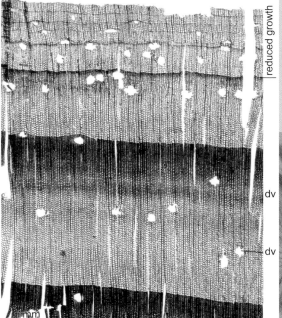

10.45 Growth reduction in Scots Pine (Pinus sylvestris) that almost completely lost their needles due to gamma radiation from the exploded nuclear power station Chernobyl.

10 Modifications Caused by Extreme Events

165

CROWN DESTRUCTION DUE TO GRAZING

In large areas of the globe, the vegetation in general, but especially tree species, are affected by grazing. The soil becomes eroded, and the surviving trees with their typical "grazing forms" remain stunted, whilst their wood production is minimal (10.46-10.48). Continuous grazing causes anomalous wood structures, such as wedging rings, small vessels and thin-walled fibers (10.49-10.52). Periodic grazing causes density fluctuations. Crown deformation, however, is reversible. Large rings after a suppressed period indicate reduced grazing.

10.46 A goat grazes an Argan Tree (Argania spinosa) in the Ameln Valley, Morocco (photo: Hänsel).

10.47 Comparison of a pair of 30-year-old Ashes (Fraxinus excelsior) on a dry site, one of which was grazed whilst the other one grew normally. The grazed plant is 50 cm tall and multi-stemmed; the other one is 4 m tall, with a single stem. Valais, Switzerland.

10.48 Severely grazed, mixed Bosnian Pine (Pinus leucodermis) and Beech (Fagus sylvatica) site near the timberline in southern Italy. The beeches have the typical, bush-like „grazed" shape. The roots were exposed as a result of soil erosion. Mt. Pollino, Italy.

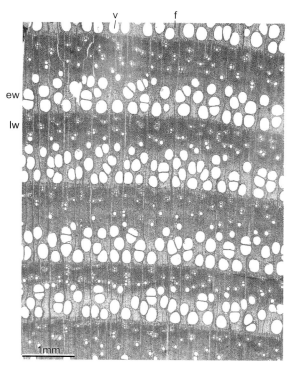

10.49 Ring width of a 15 m tall, free standing, 20-year-old Ash tree (Fraxinus excelsior), with a well-developed crown. The ring-porous wood, with big vessels and a large, dense latewood zone, is typical of this tall tree. Ticino, Switzerland.

10.50 Ring width sequence of a 50 cm tall, severely grazed, about 40-year-old Ash (Fraxinus excelsior), with a very small crown showing small vessels and almost absent latewood. Ticino, Switzerland.

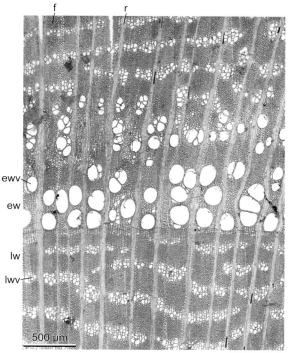

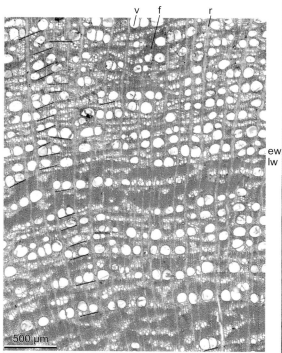

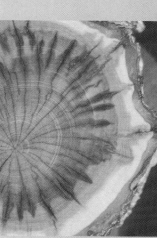

10.51 Two rings of an eight-year-old, basal Field Elm long shoot (Ulmus minor). The rings are very large and exhibit a „typical elm wood" structure. Valtellina, Italy.

10.52 Approximately 30 rings of a crippled Field Elm (Ulmus minor) by a roadside. The wood structure of this natural bonsai, with its small vessels and very narrow rings, is adapted to continuous grazing. The „typical elm wood" structure has almost disappeared. Valtellina, Italy.

10 Modifications Caused by Extreme Events

Crown Destruction Caused by Pruning and Pollarding

When branches are cut off a tree to supply animal fodder (10.53, 10.54), or for garden design purposes (10.61, 10.62), or just in order to protect buildings and roads, the photosynthetically active leaf area is reduced. Immediately after the crown has been cut, the physiological priorities of the plant change: first, wounds must be compartmentalized by chemical barriers, and new shoots must be initiated. Wood formation comes second. Therefore, pruning and pollarding reduce cell wall formation (10.55-10.57) and cause a sudden reduction in the ring width (10.58-10.60). A pruning in winter will result in a small ring in the following growing season (10.58). The effects of cutting during the growing season are reflected in the current ring by reduced lignification (10.55) and often by a false additional ring.

10.53 Recently pollarded stump of a thick Ash branch (Fraxinus excelsior). Valais, Switzerland.

10.54 Aspect of Ash trees (Fraxinus excelsior) 15 years after pollarding. The thick basal branches and straight long shoots are characteristic of these trees. Valais, Switzerland.

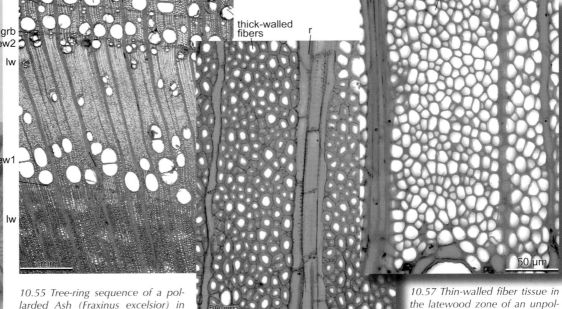

10.55 Tree-ring sequence of a pollarded Ash (Fraxinus excelsior) in August. Note the light earlywood, the narrow second ring with big vessels in the latewood (ew2), and the small rings that follow.

10.56 Thick-walled fiber tissue in the latewood zone of an unpollarded Ash (Fraxinus excelsior).

10.57 Thin-walled fiber tissue in the latewood zone of an unpollarded Ash (Fraxinus excelsior).

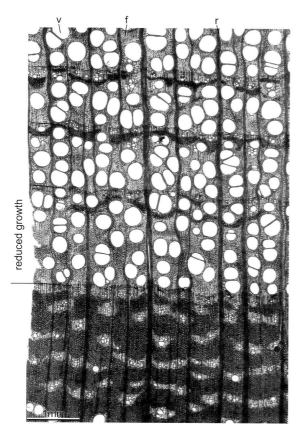

10.58 Tree-ring sequence of a pruned Honey Locust (Gleditsia triacanthos) in the city of Basle. After pruning, the rings are very narrow and almost without latewood.

10.59 Tree-ring sequence of a pruned Barberry (Berberis sp.) shrub in a hedge, showing a sudden reduction of ring and latewood width. Zürich, Switzerland.

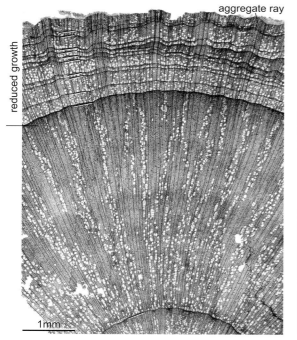

10.60 Tree-ring sequence of a pruned Hazelnut shrub (Corylus avellana) in a hedge, showing the sudden reduction of ring and latewood widths. Zürich, Switzerland.

10.61 Pruned Box Wood (Buxus sempervirens) hedge in Paris.

10.62 Pruned Horse Chestnut (Aesculus hippocastanum) in a restaurant garden. Zürich, Switzerland.

10 MODIFICATIONS CAUSED BY EXTREME EVENTS

THE FELLING OF STEMS

Many tree species even survive the removal of the entire stem and crown. Live stumps clearly show this phenomenon (10.63-10.65). A tree that survives being felled can do so only because the roots of several plants have anastomosed, i.e. they have grown together (Bormann 1962; 10.66, 10.67). After the stem has been removed, the still living cambium starts overgrowing the stump (10.63, 10.68). The same mechanisms are activated as after pruning and pollarding, but the physiological and anatomical changes are much more severe. Here, the reactions of a conifer (10.68-10.71) and a broadleaf tree (10.72, 10.73) are shown. Both trees were felled during forestry operations. The anatomical differences between the tree period and the stump period are very obvious: due to the lack of regulating hormones from the buds, the classical longitudinal tissue orientation is lost, and the tangential fiber orientation is chaotic (10.71, 10.73). The wound margin initiated a classical overgrowing process (10.68). Many species are able to produce new shoots (10.64, 10.65), but some trees survive as live stumps (10.63).

10.63 Rotten Norway Spruce stump (Picea abies) with a live outer edge. The felled tree had root contact with a live neighbouring tree. This tree's root system supplies the live stump with assimilates.

10.64 Beech stump (Fagus sylvatica) with an adventitious shoot growing out of the cambium.

10.65 Horse Chestnut stump (Aesculus hippocastanum) with an adventitious shoot growing out of the cambium.

10.66 The root systems of two Silver Firs (Abies alba) are connected (anastomosed) and leads to a new crown formation out of the stump.

10.67 Two anastomosed Norway Spruce trees (Picea abies).

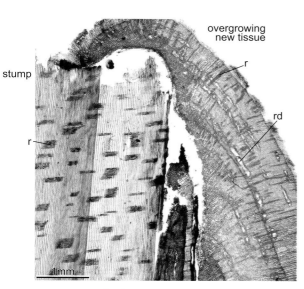

10.68 Longitudinal section of an overgrowing Norway Spruce stump (Picea abies). The dark, rotten, outer edge of the old tree has been overgrown. Note the intensive growth covering the stump.

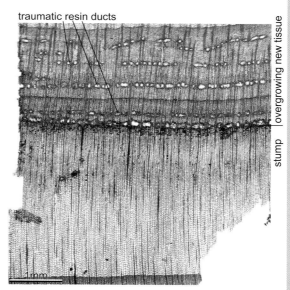

10.69 Cross-section of an overgrown Norway Spruce stump (Picea abies). The „tree rings" are large and light. The „stump rings" are small, dense and full of traumatic resin ducts.

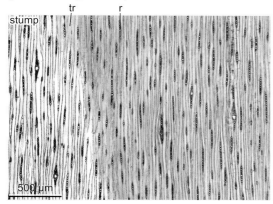

10.70 Tangential section of the „tree wood" of an overgrown Norway Spruce stump (Picea abies). Note the linear-oriented tissue.

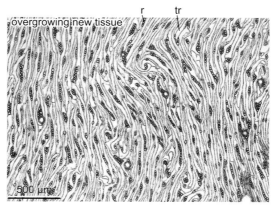

10.71 Tangential section of the „stump wood" of an overgrown Norway Spruce stump (Picea abies). Note the irregularly-oriented tissue.

10.72 Cross-section of an overgrown Beech stump (Fagus sylvatica). The „stem wood" with large rings is almost rotten. The wood is characterized by a dominance of vessels. The „stump wood" is dominated by parenchyma cells; there are virtually no vessels.

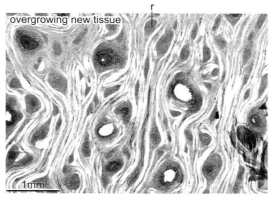

10.73 Tangential section of the „stump wood" of an overgrown Beech stump (Fagus sylvatica). Note the irregularly-oriented tissue.

10 MODIFICATIONS CAUSED BY EXTREME EVENTS

Growing Together: Anastomosis

The tissue of shoots, branches and roots of the same tree, and roots of different individuals of the same species, may merge (10.74-10.76). This capability is used in horticulture and fruit production to graft shoots (10.78) of different economic value: as soon as two live, compatible parenchyma cells touch one another, a merging process starts, leading to the formation of new callus cells (10.77, 10.79).

The strength of tree crowns is based on the natural anastomosis of two neighbouring branches (10.74); and the physical stability and sustainability of stands is based on the anastomosis of root systems (10.66, 10.67).

10.74 Two anastomosed Cork Oak branches (Quercus suber). Algarve, Portugal.

10.75 Intensively anastomosed shoot system of the Common Ivy (Hedera helix). Zürich, Switzerland.

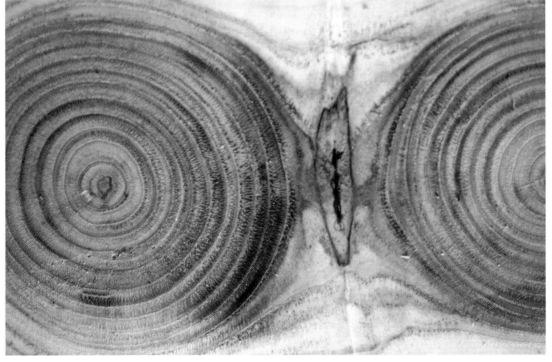

10.76 Two anastomosed Sweet Chestnut branches (Castanea sativa). Cells that touch each other stimulate growth and stabilize the connection between the two branches. Very often, some bark particles remain between the merged branches. Ticino, Switzerland.

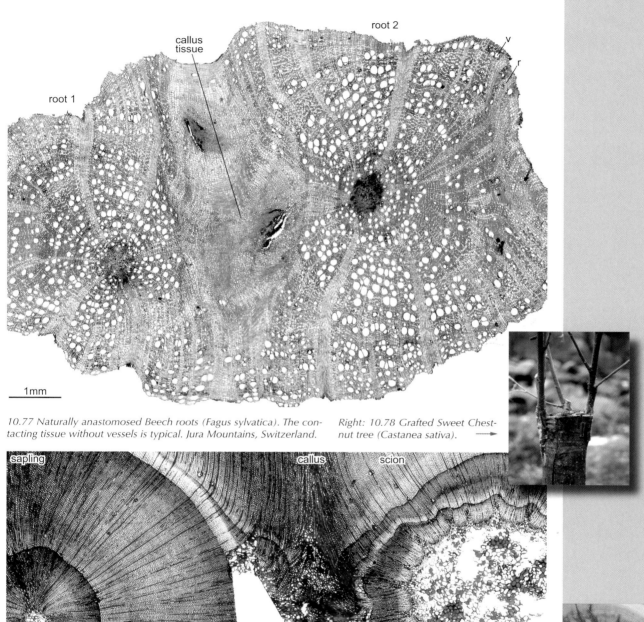

10.77 Naturally anastomosed Beech roots (Fagus sylvatica). The contacting tissue without vessels is typical. Jura Mountains, Switzerland.

Right: 10.78 Grafted Sweet Chestnut tree (Castanea sativa).

10.79 Grafted Norway Spruce (Picea abies). A shoot with a big pith has been grafted onto a sapling with a small pith. Growth increased after the tissue had grown together. Tree nursery, Zürich, Switzerland.

10 Modifications Caused by Extreme Events

Crown, Stem and Site Destruction by Forest Fires

Fire is a natural phenomenon in most forest ecosystems (Goldammer and Furyaev 1996; 10.80). It influences tree growth, longevity as well as the composition of the vegetation (10.81-10.83). Here, we discuss effects of fire on tree individuals. Throughout history, vegetation and trees have adapted to fire. Certain species have acquired properties that protect them from fire, e.g. by the formation of thick, fire-resistant bark (10.89) or protective leaf base layer (10.87), the ability to create adventitious shoots at the root collar (10.84, 10.88) or at the stem (10.86). Cones of some species release seeds under the influence of heat (10.85).

10.80 Forest fire in the Canadian boreal conifer zone with fire fronts and conspicuous smoke columns.

10.81 The patchy pattern of a forest fire in the Canadian boreal conifer zone. The white spots show the area where the thin humus layer has been burned to the mineral soil. The black parts indicate zones where tree stems are not burnt but carbonized. In the light brown zones only the needles of the trees have been burnt. Characteristic is the mosaic between affected and unaffected (green) zones.

10.82 Regeneration of coniferous forest after a fire in the Canadian Rocky Mountains. Twenty years after the fire a new generation of fir grows between standing dead trees.

10.83 Effect of a crown fire on Grass Trees (Xanthorrhoea sp.) and on Eucalypts (Mallee). Western Australia.

10.84 A fire killed the shoots above ground of a Broom (Genista sp.) and stimulated the formation of adventitious shoots. Pyrenees, Spain.

10.85 Cones of Monterey Pine (Pinus radiata). Cones remain closed under normal climatic conditions (above), they open their scales under the influence of the heat of fire (below). Plantation, Southern Australia.

10.86 Two-year-old shoots on a stem of a Canary Pine (Pinus canariensis). The ability to form adventitious shoots on stems is an adaptation to the occurrence of frequent fires. La Palma, Canary Islands.

10.87 This Grass Tree (Xanthorrhoea sp.) survived a fire that occurred two months before. The tips of the leaves were still affected by the fire, but the meristem itself was well protected by a thick layer of old leaf bases. Booderee Ntl. Park, Western Australia.

10.88 Vegetative regeneration after a fire. The root stock survived and formed new sprouts. Sweet Chestnut (Castanea sativa), Ticino, Switzerland.

10.89 The very thick bark of Araucaria trees is an adaptation to the occurrence of frequent fires. The thick cork mantle around the stem is a heat barrier and prevents the cambium from damage. Lonquimai, Chile.

continued next page

10 MODIFICATIONS CAUSED BY EXTREME EVENTS

Fire scars are very conspicuous (10.90-10.94). The consequence of crown destructions is an interrupted or reduced growth. Reduced photosynthetic activity and transpiration immediately stop cell wall growth, and the formation of big vessels. Therefore, intra annual bands of thin walled fibers (10.95) and small vessels (10.96) mark the occurrence of fires. Long-term reactions to fires are abrupt growth reductions, often correlated with missing rings (10.97).

Stem destructions by heat are in fact damages to the cambium. Cambial reactions to moderate heat are tangential rows of resin and gum ducts (10.98, 10.99), cell collapses, tyloses and callus cells (10.100). The result of an intensive heat is a locally destroyed cambium. The wound stimulates the formation of scars. The process is based on accelerated growth of surviving lateral cambial and parenchyma cells (10.101, 10.102). □

10.90 Ground fire in Siberian Scots Pine forest (Pinus sylvestris) destroys the cambium and causes fire scars. It burns the heartwood as in a chimney.

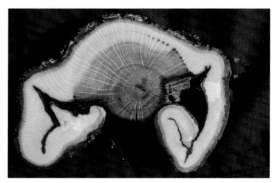

10.91 Carbonized fire scar on a Ponderosa Pine (Pinus ponderosa) tree. Many fires affected the tree at the same location.

10.92 Fire scar at the lower side of an Oak branch resulting from ground fire. Ticino, Switzerland.

10.93 Three fire scars on a stem of Dahurian Larch (Larix dahurica) in the boreal zone of Siberia. The first fire affected the tree after 44, the next after an additional period of 15 years and the third fire after additional 25 years. Eastern Siberia.

10.94 Many fire scars on a White Gum tree (Eucalyptus pauciflora) at timberline in Eastern Australia. Man made grass fires hit the tree many times, approximately nine scars can be observed at the lower and four scars at the upper side of the tree disc. Blue Mountains, Australia.

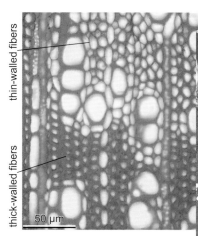
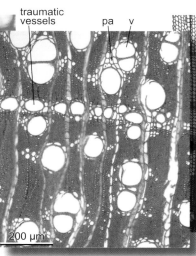
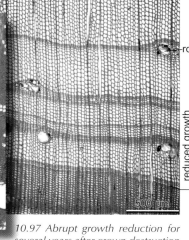

10.95 Abrupt reduction of cell wall growth after the destruction of the crown by a fire. Emu Bush (Eremophila sp., Myoporaceae), Australian Savannah, Queensland, Australia.

10.96 Tangential row of small vessels after the destruction of the crown by a fire. Acacia hotwittii, Fabaceae. Australian Savannah, Queensland.

10.97 Abrupt growth reduction for several years after crown destruction by a fire in a Ponderosa Pine tree (Pinus ponderosa), Colorado, USA (slide: Ortloff).

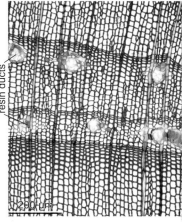
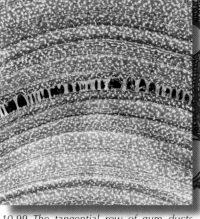
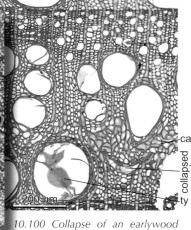

10.98 Formation of tangential rows of resin ducts in the xylem of a pine after moderate heat influence. Radial growth reduction is probably a reaction to crown destructions. Scots Pine (Pinus sylvestris), Siberia.

10.99 The tangential row of gum ducts (kino veins) in a Eucalypt stem is a reaction to moderate heat of a forest fire. Australia.

10.100 Collapse of an earlywood vessel, tyloses and callus formation in the xylem of an oak after moderate heat influence. Downy Oak (Quercus pubescens), Valais Switzerland.

Left: 10.101 Fire scar on the stem of a tropical Proteaceae. Characteristic is the vessel-free zone beside the wound and the intensive growth towards the wound. Queensland Australia.

Right: 10.102 Hidden fire scar on the stem of a Ponderosa pine (Pinus ponderosa). The zone with unlignified cells indicates that the fire occurred late in the growing season. Surviving zones over the wound (above the crack) first produced callus cells. At the end of the normalization process a row of tangential resin ducts was formed. Boulder, Colorado, USA.

Crown and Stem Destruction By Parasites and Pathogens

Trees are host to an immense diversity of parasites (10.103). Those feeding on leaves reduce growth and cause anatomical changes in the stem. The consequences of parasites that feed on the stem vary. Mistletoe haustoria penetrate live xylem without triggering any intensive defence reaction (10.104-10.107). Increased formation of resin ducts is a reaction to witches' broom on conifers. Many cancers initiate callus and heartwood formation or disturb the water-flow. Here, a small selection of often observed reactions to mistletoes, Chestnut Blight (10.108-10.111) and Dutch Elm Disease (10.112-10.114) is shown.

Right: 10.103 Aspect of White-berried Mistletoe (Viscum album) on a Scots Pine twig (Pinus sylvestris). Valais, Switzerland.

10.104 Cross-section of a Scots Pine stem (Pinus sylvestris) with a White-berried Mistletoe stem (Viscum album) and haustoria. Some haustoria are dead and overgrown. Valais, Switzerland.

10.105 White-berried Mistletoe haustoria (Viscum album) on a Scots Pine twig (Pinus sylvestris). They became established within two years. The host doesn't show any defense reactions. Valais, Switzerland.

10.106 Cross-section of haustoria from White-berried Mistletoe on a Sycamore twig (Viscum album on Acer pseudoplatanus). The unlignified haustoria are overgrown by the lignified xylem (slide: Zibulski; polarized light).

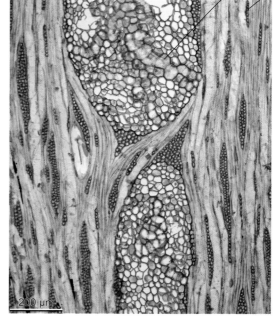

10.107 Tangential section of haustoria from White-berried Mistletoe on a Sycamore twig (Viscum album on Acer pseudoplatanus). The fibers of the xylem tissue grow around the haustorium (slide: Zibulkski).

Left: 10.108 Sweet Chestnuts (Castanea sativa) affected by Chestnut Blight (Cryphonectria parasitica). Most trees died, but this one formed new shoots from the basal part of the stem. Ticino, Switzerland.

Left: 10.109 Cross-section of a Sessile Oak branch (Quercus petraea) The hidden scars show that the tree survived three Chestnut Blight attacks (Cryphonectria parasitica). Ticino, Switzerland (material: Heiniger).

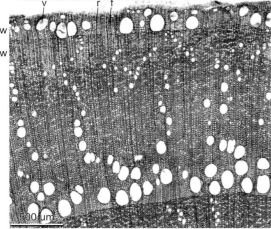

10.110 Dead Sweet Chestnut twig (Castanea sativa). Chestnut Blight (Cryphonectria parasitica) killed the plant in spring, just after the first earlywood vessels had been formed.

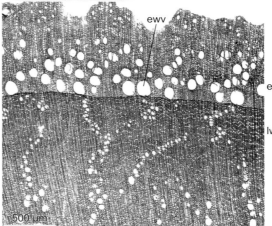

10.111 Dead Sweet Chestnut twig (Castanea sativa). Chestnut Blight (Cryphonectria parasitica) killed the plant in early summer, after latewood formation had begun.

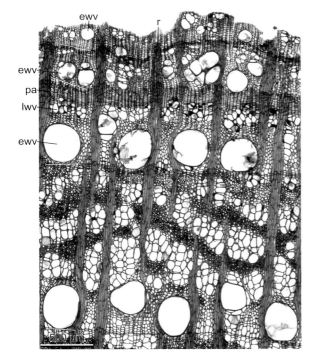

10.113 Dead Wych Elm twig (Ulmus glabra). Dutch Elm Disease (Ceratocystis ulmi) infected the plant two growing seasons before death, and killed it in spring, just after it had formed the first earlywood vessels. The moment of the infection can be determined on the basis of the large parenchymatous latewood.

Right: 10.112 Dutch Elm Disease (Ceratocystis ulmi) in an elm stand. Many trees died, some survived by forming new shoots. Lund, Sweden.

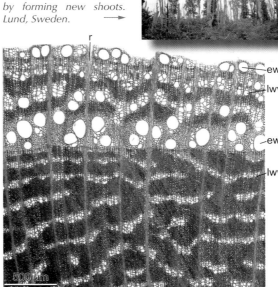

10.114 Dead Wych Elm twig (Ulmus glabra). Dutch Elm Disease (Ceratocystis ulmi) infected the plant in the growing season before death, and killed it in the spring, just after it had formed the first earlywood vessels. The moment of the infection can be dated on the basis of the reduced latewood zone.

10 MODIFICATIONS CAUSED BY EXTREME EVENTS

Mechanical Stress on Stems due to Imbalance and Shock

Growing processes in the crown and root system, as well as geophysical processes, create imbalances in plants (10.115). As a result, reaction wood forms. First, tree stems become eccentric by forming more cells on the stressed side: conifers produce compression wood (10.117, 10.119) and many angiosperms form tension wood (10.116, 10.118, 10.120, 10.121). Periods of stress in trees may be reconstructed by dating and analyzing the beginning and duration of reaction wood formation (10.117-10.120) and its direction (Timell 1986; 10.122).

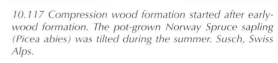

Left: 10.115 Green Alder (Alnus viridis) in an avalanche channel. The bent stems have tension wood on the upper side. Susch, Swiss Alps.

10.116 Tension wood formation at the beginning of the earlywood indicates an event that took place in winter. An avalanche tilted this Silver Birch (Betula pendula). Susch, Swiss Alps.

10.117 Compression wood formation started after earlywood formation. The pot-grown Norway Spruce sapling (Picea abies) was tilted during the summer. Susch, Swiss Alps.

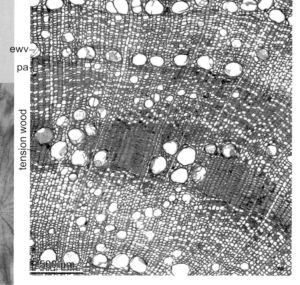

10.118 Tension wood formed for one year in a suppressed Sweet Chestnut tree (Castanea sativa). It probably was a reaction to being bent over by snow for a short time. Zürich, Switzerland.

10.119 Compression wood formed for seven years in a Mountain Pine branch (Pinus mugo). Tschierv, Switzerland.

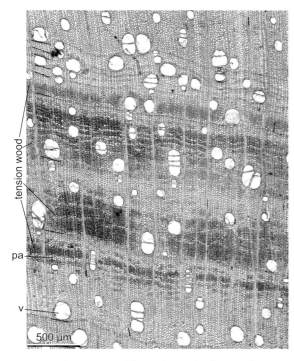

10.120 Tension wood formed periodically in a Common Walnut branch (Juglans regia). This probably was a response to strong wind. Zürich, Switzerland.

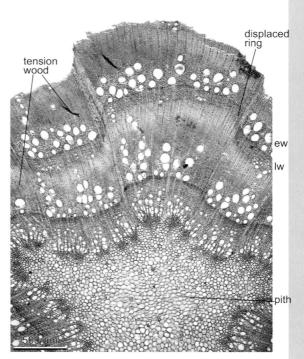

10.121 Tension wood formed in slightly different directions in a Common Oak twig (Quercus robur). Valais, Switzerland.

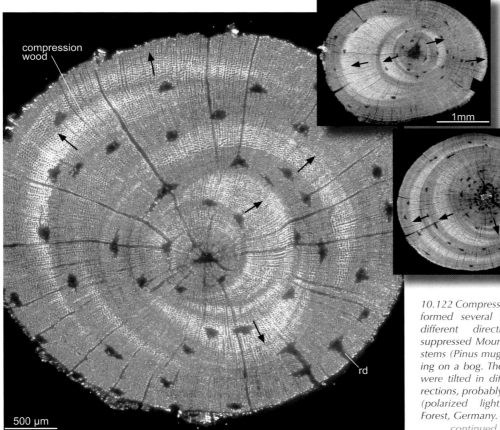

10.122 Compression wood formed several times, in different directions, in suppressed Mountain Pine stems (Pinus mugo), growing on a bog. The saplings were tilted in different directions, probably by snow (polarized light). Black Forest, Germany.

continued next page

10 Modifications Caused by Extreme Events

10 Modifications Caused by Extreme Events

Rock fall (10.123) primarily causes crushed cell walls (10.124, 10.125). When stems bend over because of extremely strong winds (10.126), tangential cracks form in the earlywood. The result are pitch pockets (10.127, 10.128).

10.123 A stone hit a Scots Pine (Pinus sylvestris) and got stuck in the stem. The shock crushed cell walls inside the stem. Ticino, Switzerland.

10.124 Zone of crushed tracheids in a Norway Spruce stem (Picea abies; polarized light) after a rockfall. Research plot of the Institute of Snow and Avalanche Research, Davos.

10.125 Crushed tracheid cell walls after a rock fall. Norway Spruce (Picea abies; polarized light). Research plot of the Institute of Snow and Avalanche Research, Davos.

10.126 Wind-exposed Mountain Hemlock (Tsuga mertensiana) at the upper timberline. The stems contain resin pockets. Crater Lake, Oregon, USA.

10 Modifications Caused by Extreme Events

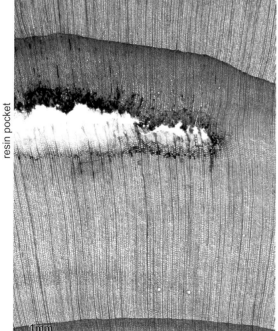

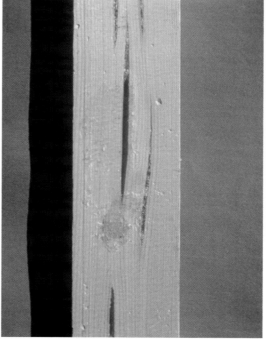

10.127 Resin pocket in a Norway Spruce stem (Picea abies). Severe stem-bending by strong winds caused a tangential crack. Later on, the crack expanded and filled with resin.

10.128 Longitudinal section through resin pockets in a Norway Spruce board (Picea abies).

Physiological Stress Caused by Stem Wounds

Stems are constantly exposed to physical and biological factors that may damage them. Plants are able to protect themselves by producing chemical barriers and overgrowing zones around wounds (see also p. 62). With the dendrochronological dating of injuries, scars have become important witnesses of the reconstruction of geomorphological events (Schweingruber 1996).

Scars caused by avalanches (10.115), rockfall (10.123), landslides, storms, hail (10.129-10.131) and erosion (10.132) are frequent in mountain regions. Biological damage from insects and vertebrates occurs all over the world. Here, only tooth marks from mice (10.133, 10.134), and deer (10.135) trampling damage by deer and man (10.136), and peck marks by woodpeckers 10.137-10.139) will be mentioned.

Right: 10.129 Hail scars on a young, Sessile Oak twig (Quercus petraea). Arth Goldau, Switzerland.

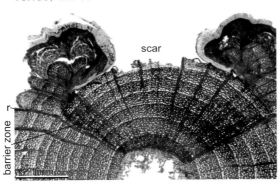

10.130 Cross-section of a hail scar on a Rock Currant twig (Ribes petraeum). The barrier zones, along the rays and the overgrowing zones, with a distinct periderm, are conspicious. Arth Goldau, Switzerland.

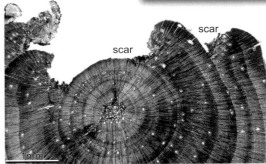

10.131 Cross-section of a Mountain Pine twig (Pinus mugo) with two hail scars. Both events occurred at the beginning of the growing season. Arth Goldau, Switzerland.

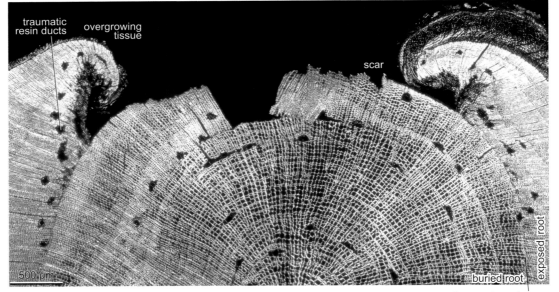

10.132 Cross-section of an exposed and injured Mountain Pine root (Pinus mugo) on a river shore. When the root was exposed, structural changes took place, whilst the overgrowing zone dates the injury (polarized light). Tschierv, Switzerland.

Left: 10.133 Small Beech stem (Fagus sylvatica) with a mouse scar. Zürich, Switzerland.

Right: 10.135 Deer scars on a Cottonwood stem (Populus sp.). Netherland, Colorado, USA.

10.134 Cross-section of a small Ash stem (Fraxinus excelsior) with a mouse scar. The injury occurred after the first year of life, and the plant easily survived it. The wounded section is compartmentalized, and the remaining part is completely healthy. Zürich, Switzerland.

10.136 Cross-section of an exposed Mountain Pine root (Pinus mugo) on a deer track. Scars and callus tissue indicate the trampling frequency. Tschierv, Switzerland.

Right: 10.137 Peck marks on Scots Pine (Pinus sylvestris) are a result of repeated visits of sap-sucking woodpeckers. Vienna, Austria (material: Wimmer).

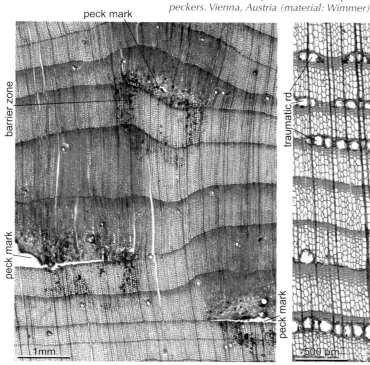

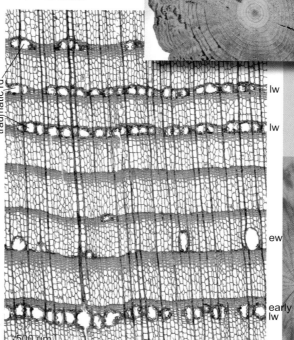

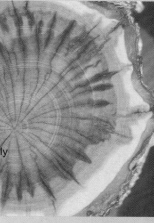

10.138 Three single peck marks of a sap-sucking woodpecker on a Scots Pine stem (Pinus sylvestris). The woodpecker always attacked the tree at the beginning of the growing season. Valais, Switzerland.

10.139 Repeated rows of tangential, traumatic resin ducts in Dahurian Larch (Larix dahurica) from the activity of sap-sucking woodpeckers. They visited the tree at the beginning and the end of the growing season. Eastern Siberia.

CHAPTER 11

FROM ANATOMICAL FEATURES TO PLANT STRUCTURES

The following chapter will explore the extent to which anatomical features are used to produce the large variety of the observed whole-plant structures in nature.

Former dominant canopy tree with tall stem, broad crown, and horizontal branches. Kapok Tree (Ceiba pentandra), Costa Rica.

How Do Woody Plants Get Old?

As all organisms, woody plants undergo senescens and aging at the cellular level. Phloem cells may live for up to three years, xylem cells function a few decades, but trees may live centuries to millennia, and plants producing rhizomes may live forever. Therefore, there are special features to get old:

(i) An active cambium which continues to maintain the essential processes of life despite permanent senescence of individual cells.

(ii) The maintenance of a positive carbon balance. Since plants are not mobile, the space for leaves is set by the area the plant can reach. Thus, the relation between assimilating leaves and respiring stems and roots decreases with the size of the plant, but it must be maintained in favor of assimilation also at increasing size and age. Survival can be achieved through die-back, and by heartwood formation within the tree stem (11.1). Die-back takes place in desert shrubs where part of the multi-stem shrubs dies under adverse conditions, and few shoots remain alive. The most conspicuous die-back is found in trees, where the cambium dies back in a major part of the circumference (*Pinus longaeva*, 11.4). An alternative mode of die-back is the formation of rhizomes. The creeping stem of a rhizome continues to form new shoots at the apex and the organism dies back at the opposite end. Under the special conditions of bogs (11.2, 11.3) or of permafrost (11.5, 11.6), the dead plant material is conserved from decomposition and is witness for the time since germination.

(iii) The prevention of decomposition through production of toxic substances. Most woody plants produce toxic compounds to avoid attack by fungi which are specialized to break down wood (11.7).

Right: 11.1 Syrian Plum (Prunus domestica ssp. syriaca): the sapwood contains dead vessels for water conduction and living parenchyma for storing carbohydrates and nitrogen-compounds. The dark center of the cross section is void of any living tree cells. The wood is impregnated by the alkaloid Cytisine to avoid decomposition and is colored by secondary plant compounds. →

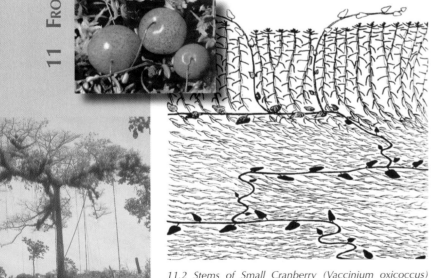

11.2 Stems of Small Cranberry (*Vaccinium oxicoccus*) growing on bogs die due to lack of oxygen under the wet conditions. The plant continues to grow by formation of new rhizomes which are able to compete for light with the growing Sphagnum cover. Individual plants may have germinated when the bog started to grow, which, in Central Europe, was presumably about 10,000 years ago (after Walter 1968). Small left: from Aeschimann et al. 2004.

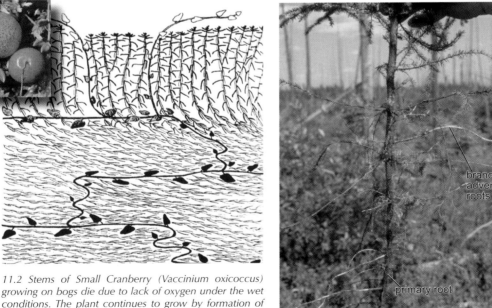

11.3 Young Black Spruce tree (*Picea mariana*) uprooted from a bog. The stem was buried in the growing Sphagnum layer, and adventitious roots developed on the buried branches. Quebec, Canada.

11.4 The oldest known living tree species, Bristlecone Pine (Pinus longaeva), may reach an age of 5000 years. It grows in the alpine deserts of Nevada and California at about 3500-4000 m elevation. Under the cold and arid conditions wood decomposition is slow. The tree maintains a positive carbon balance by adjustment of the girth of the active cambium. Most of the circumference may die and only a stripe of cambium survives. Thus, the tree continues to grow like a board. Old trees may be 2 m wide and 0.2 m broad. Most parts of the conspicuous old trees are, in fact, dead.

11.5 Detail from the Duvany Yar loess wall: Bone of a mammoth that was, like the plant rhizomes, buried by the loess.

11.6 At the Duvany Yar site in Northeast Siberia, a wall of loess, about 100 m high is exposed by the Kolima river. Loess deposition started about 1 million years ago, and the vegetation continued to grow with the increasing loess layer. At the same time, the permafrost rises at the rate of loess deposition, it conserves the lower plant parts like in a freezer.

11.7 Cross section of a 760-year-old Western Red Cedar (Thuja plicata). Thuja may survive to be more than 1000 years old, maintaining an active cambium around the stem. The sap-wood is hardly visible at the outside. Despite the protection by Thujaplicin, a large part of the cross section has been attacked by fungi. Pacific Northwest, USA.

How Large Can Trees Get?

With the invention of apoptosis and heartwood formation, woody species can in principle form individuals of any size and shape. The heartwood is a non-respiring structural dead biomass which serves only to maintain the tree structure. The size is eventually only limited by the physical properties of wood, by the protection against wood decomposing organisms, and by the cellular capability to conduct water and carbohydrates along a large plant structure in the active phloem and xylem (Schulze et al. 2005). Tree height is obviously limited by water transport. The tension in the xylem vessels cannot exceed a certain limited range, depending on the diameter of the vessels. Although the tensile strength of a water column as measured in the laboratory could support 200-300 m tall trees, in reality trees have only reached 100-150 m (11.8). In contrast, tree diameter is basically unlimited (11.9, 11.10), and may continue to grow even if the center of the tree is taken away by fire (11.11) or rots from fungal attack (11.12, 11.13).

Below: 11.9 The thickest known tree is the Kauri (Agathis australis) on the North Island of New Zealand (7 m diameter, 50 m height without any significant taper).

Above: 11.8 The Jarrah (Eucalyptus marginata) of Southwestern Australia. At 140 m it is the tallest known tree species. It can get this tall because of its dense wood and its small vessels.

11.10 The Mammoth Tree (Sequoiadendron giganteum) reaches the largest volume among trees (5500 m³ of wood, 6 m diameter, 96 m height). Sequoia National Park, USA (photo: Etzrodt).

11.11 With repeated fires, the center of this Red Tingle Tree (Eucalyptus jacksonii) has burnt away, but the tree continues to grow along the outer circumference. New Albany, Western Australia.

11.12 The Olive Tree (Olea europea) is attacked by Polyporus fulvus, a fungus which decomposes the heartwood while the sapwood remains intact. The original tree disintegrates into many secondary stems, which continue to grow. The cultivated Olive is an anthropogenic cultivar which is propagated through grafting onto the root stock of the wild ancestor, Olea sylvestris since biblic times. Mallorca.

11.13 Many trees form very irregular branches which may support epiphytes. Even other trees may germinate and grow on top of old irregularly shaped canopy trees using the rotten tree center as substrate. An example is the Wheel Tree (Trochodendron aralioides), growing at 1000 m elevation on Yakushima Island, Japan.

The Structural Diversity of Woody Plants

The structural diversity is unlimited. Wood is similar to steel enforced concrete (cellulose enforced with lignin) which is eventually limited in size and shape by its ultimate strength. Thus, the regulation of plants structure can in principle occur in other tissues, according to the demands for light and nutrients, or in response to environmental stress or herbivores. In woody plants the structure is generally regulated by the dominance of the bud (11.14, Schulze et al. 2005). It leads to the formation of trees and shrubs (11.15-11.18) or of herbaceous species (11.19). Thus, woody plants range from tall canopy trees to herbaceous species which may maintain a woody anatomy in the hypocotyls and in the root. In addition, the above ground stem may not always be of solid wood. There are herbaceous stems, e.g. of *Coreopsis*, there are succulent trees of the Cactaceae and the Euphorbiaceae (11.21-11.26), and woody below-ground stems, e.g. of *Anisum sativum* (11.20). There is no systematic design to tree structures; in fact, everything seems possible. Structure depends only on the physical properties of tension and compression of wood and on the demand by buds which are responsible for the shape of the crown as we recognize it in the field.

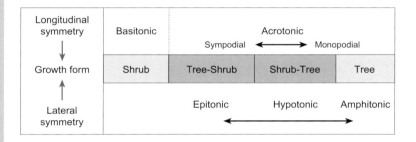

Left and below: 11.14 The activity of buds determines the latitudinal and longitudinal symmetry of woody plants (Schulze et al. 2005). The bud activity along the longitudinal axis determines whether the resulting structure is a tree or a shrub.

	Rubus	*Rosa*	*Ribes*	*Prunus*	*Crataegus*	*Cornus*	*Acer*	*Fagus*
Branching pattern	Sympodial	Sympodial	Sympodial	Sympodial	Sympodial	Monopodial	Monopodial	Monopodial
Longitudinal symmetry	Basitonic	Basitonic	Weakly basitonic	Mesotonic acrotonic	Mesotonic acrotonic	Mesotonic acrotonic	Mesotonic acrotonic	Mesotonic acrotonic
Lateral symmetry	Strongly epitonic	Strongly epitonic	Strongly epitonic	Weakly epitonic	Epitonic and hypotonic	Weakly epitonic	Strongly hypotonic	Amphitonic

11.15 The arrangement of woody plants in a hedgerow basically represents a sequence of woody plants changing from basitonic and epitonic to more acrotonic and amphitonic bud dominance. The edge of a hedgerow is occupied by Blackthorn (Prunus spinosa) as woody pioneer species, followed by Blackberry (Rubus sp.) and Rose (Rosa sp.) species. Towards the center, Hawthorn (Crataegus sp.) and finally Buckthorn (Rhamnus cathartica) and Common Maple (Acer campestre) follow. Stadtsteinach, Germany.

11.16 Norway Spruce (Picea abies). In most coniferous trees the apical dominant bud remains the most active bud (acrotonic).

11.17 With the more active bud on the upper side of a branch (epitonic), Blackberry (Rubus sp.) produces annual bowing shoots which may reach several meters (above). They can form roots again at the apical end (below).

11.18 A dominant bud at the lower side of a branch (hypotonic) results in a wide ranging branch system, as found, e.g. in Common Maple (Acer campestre).

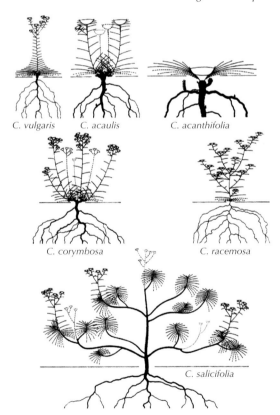

11.19 Carline Thistles (Carlina sp.): under non-favorable growing conditions, especially drought, the terminal buds do not maintain growth as in C. salicifolia from Teneriffe. They change into a woody stem of the hypocotyls which develops only a terminal rosette as in C. acanthifolia, growing in the northern Mediterranean region (after Meusel 1976).

11.20 Anise (Anisum sativum), growing in the Mediterranean, forms a massive stem in the soil which carries a rosette of leaves at the soil surface. El Palmar, Tenerife.

continued next page

11 FROM ANATOMICAL FEATURES TO PLANT STURCTURES

11.21 Cacti can grow up to sizable trees. Cardon (Lemaireocereus weberi) growing in the highlands of Central Mexico builds 15 to 20 m structures composed of branching succulent stems.

11.22 The Kokerboom of Namibia (Aloe dichotoma) is a monocot which forms trees where the stem carries an apical rosette. When the rosette flowers, the apical meristem is used, and the tree continues to grow from two buds formed at the base of the rosette. This results in a dichotomous branching pattern.

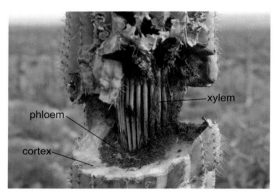

11.23 The fleshy ridge of the cactus column is linked to the vascular bundle in the center. The succulent structure can swell and shrink like an accordion, and thus it can store water. Saguaro (Carnegiea gigantea), Arizona, USA.

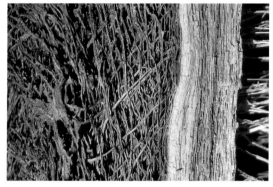

11.24 The stem of Aloe dichotoma is formed by a solid tube of the bark, the parenchyma and the outer vascular bundles. The center is composed of single strands of closed vascular bundles which are embedded in a parenchyma.

11.25 Neoraimondia arequipensis is a tree-like cactus growing at the foggy cold coast-line of Peru.

Below: 11.26 The leathery epidermis prevents water loss during the day when the plant decarboxylates organic acids which are produced overnight when the stomata are open.

Protection Against Environmental Extremes
Temperature Extremes

Hot as well as cold temperature extremes are found near the soil surface. Special structures allow plants either to tolerate or to avoid these extremes. In subtropical hot deserts with high solar radiation Acacias form a broad and dense umbrella-shaped crown and thus shade the hypocotyls which penetrate the soil surface (11.27). If the plant cannot produce such a dense canopy, an extended height growth of antenna-like shoots allows the plant to develop leaves above the surface boundary layer with its extreme temperatures (11.28).

11.27 Shading the trunk by the umbrella-shaped crown of Red Umbrella Acacia (Acacia reficiens). This is a common protection of the base in subtropical savannahs. Namibia.

11.28 In contrast, the antenna-like growth of the Blackthorn Acacia (Acacia mellifera) avoids the temperature extremes of the near boundary layer. Nevertheless this requires shading by shrub-like lower branches. Namibia.

Extended stem growth also protects rosette plants under tropical alpine conditions, where, every night, temperatures may drop below freezing point. Thus, a forest-like vegetation is found above the „normal" tree line (11.29-11.32). Some of the stems are hollow and filled with circulating water to avoid freezing (11.30).

11.29 Senecio keniodendron has a tree-like structure carrying a large leaf rosette. The high water content of the pith in combination with the coat of dead leaves keeps the cambium from freezing also at low temperatures over night. Mount Kenya.

11.30 In Lobelia telekii the center of the stem is hollow and filled with a column of water which extends into the root. Due to the temperature gradient, a gravity driven circulation of the water provides sufficient heat to avoid freezing of the shoot and of the flowers, which are additionally protected against freezing by long bracts. Mount Kenya.

continued next page

11.31 The rosette trees are found on all continents, like here in the winter-rain regions of Africa (Pachypodium namaquensis). Richtersveld, South Africa.

11.32 The stem of Australian Grass Trees (Xanthorrhoea sp.) is similar to the parenchyma rich stems of other monocots. The stem is protected against fire by the old leaf bases, which produce a red resin, the „dragon blood", which does not burn.

The opposite strategy is found in cushion plants. In alpine climates these plant forms use the heat that is generated near the ground by solar radiation. In addition, the cushion traps the CO_2 from soil respiration and thus compensates for the decrease in atmospheric CO_2 at high elevation. Also, within the cushion, air humidity is higher than outside, which favors stomatal opening. Any branch which grows beyond the common smooth surface experiences the harsh zonal climate which will reduce its growth. Thus, the cushion plant is an example where the individual shoot is constrained by the activity of the shoot community (11.33-11.35). □

Right: 11.33 Yareta (Azorella compacta) is a shrub growing at about 5000 m elevation in the Andes of Peru. The smooth surface protects the plant from cold wind, and solar radiation heats the plant surface and traps respiratory CO_2 of stems, roots and decomposers. →

11.34 In alpine environments, conditions in winter are more favorable below the snow, where temperatures are near freezing point, and penetrating solar radiation may even allow photosynthesis. This leads to the formation of cushion-like shrubs in species which normally would grow as trees, such as Bristlecone Pine (Pinus longaeva) at the alpine tree line in Nevada.

11.35 Cushion plants are also found in coastal environments. Limonium caprariense, growing on the rock surface towards the ocean at Formentor, Mallorca, forms a dense rosette which gives the least aerodynamic resistance against splashes of salt water. Any shoot that extends beyond the surface may get a higher load of salt spray.

Protection Against Environmental Extremes
Avoiding Shade

Lianas are known to have no supporting structure but use support by other trees, and invest in length growth in order to reach light in dense canopies (11.36, 11.37). This requires enormous tensile strength, such as in steel ropes, and special modifications in the xylem in order to supply a very large plant structure with water. Thus, we find the largest vessels and the highest volume flow in lianas as compared to any other plant (Schulze *et al.* 2005).

11.36 Some lianas may improve their nitrogen supply by catching insects. The pitcher plant Nepenthes is a liana. Bogor, Indonesia.

11.37 Lianas are a common life form in dense secondary growth of tropical forests. Most lianas cannot climb, but grow with the canopy of the trees. Thus the liana is lifted from the ground by rapid apical growth. Manú Ntl. Park, Peru (photo: Dietz).

Epiphytic growth is another strategy to avoid shade in a dense forest. Fig trees are specialists which make use of the light environment of the forest canopy. The seeds are carried by birds onto canopy branches, and the seedling grows initially as epiphyte, which sends a long root straight through the canopy (11.38). After reaching the ground, and thus securing the water balance of the plant, *Ficus* produces additional roots. The stem of the host is enmeshed over time by a network of secondary roots of the fig (11.40). This kills the host, because the sugar transport of the host to its roots is interrupted. The fig exploits the resources which are released from the decay of the host-wood. At the end, a chimney-like hole reminds one of the host (11.39), while the fig continues to produce stilt roots. □

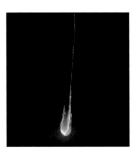
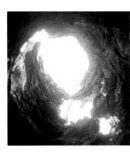

11.38 The beginning: the fig seedling sends the first root down to the forest floor.

11.39 The end: only a hole inside the fig root network remains of the host tree.

11.40 Huge Strangler Fig tree (Ficus obliqua) on Fraser Island, Australia. The host tree inside the fig root mesh is already dead and decaying.

Protection Against Environmental Extremes
Storage of Reserves in Seasonal Climates

Seasonal climates require the formation of reserves which are generally stored in the ray cells of the wood. However, some tree species also form a storage parenchyma, which in turn changes the plant structure. Most conspicuous are the „bottle" trees of the subtropical savannah region (11.41-11.43). The same principle is found in desert shrubs (11.44). The „bottles" do not serve as water store. The tree structure would not allow to expand and to shrink as in cacti. However, the „bottle" stores reserves, such as carbohydrates and amino-acids, and therefore the bottle trees of Africa are heavily damaged by hungry elephants in the dry season (11.42). Also the underground buttress of the Mallee Eucalypt of Australia may not only protect the plant, but also serve as storage organ to regenerate the shoots after burning (11.45).

11.41 Bottle-trees are found in most subtropical savannahs, e.g. the Australian Baobab (Adansonia gregorii), Bombacaceae, which may reach 14 m circumference.

11.42 Due to the soft and nutritious parenchyma, the African baobab (Adansonia digitata) is usually heavily damaged by elephants.

11.43 The Queensland Bottle Tree (Brachychiton rupestre), Sterculiaceae, forms similar „bottles" in the summer rainfall areas of Australia.

11.44 The same strategy is followed by numerous shrubs, such as the Rock Corkwood (Commiphora saxicola), Commiphoraceae, in Namibia.

11.45 The buttresses formed by the Mallee Eucalypt (Eucalyptus socialis) serve also as storage organ to allow re-sprouting after fire. Broken Hill, Australia.

Other Special Ecological Adaptations
Herbivory and Ant Plants

Protection against herbivory has led to a diverse range in plant structures. Under most conditions thorns are formed from specific branching patterns as a result of differential bud activity (see 11.14, p. 192). In some species, however, protection against herbivory originates from modifications of the stipules which are formed into spines (11.46-11.48) or by special structures which serve as nest for symbiotic ants (11.49, 11.50).

Right: 11.46 The surface of Anthyllis fulgurans growing on Mallorca is covered by a network of thorns; leaves grow only in a layer below the thorns, which sheep and goat cannot reach.

11.47 In the savannah of East Africa, in Acacia drepanolobium the two stipular spines merge and form a bulb-like structure with two openings for the ants to enter. With wind, these openings act like a recorder („Flute Acacia").

11.48 In Acacia cymbispina of Mexico, the stipules are initially formed as woody thorn-like structures. If the margins of the leaf-like stipule grow together, a hollow spine is formed which is ideal for ants to live in.

11.49 In Cecropiaceae, ant colonies live in the hollow stem; they reach the outside via pre-formed holes on the stem wall. Rancho Grande, Venezuela.

11.50 Co-evolution with ants goes further in Myrmecodia sp. (Rubiaceae), where the hypocotyl grows as a bulb-like structure which contains open channels and cavities for the symbiontic ant to nest. The plant is not only protected against herbivores, it also gets additional nitrogen from the excretions of the carnivorous ant. New Guinea.

Other Special Ecological Adaptations
Mangroves and Flooding

Mangroves represent the border between the ocean and land in tropical climates along muddy shore lines (11.51). Mangroves are a functional group of species which are able to grow in salt water by either excluding the entry of salt during water uptake or by excreting salt from special glands. The main problem of mangroves is the uptake of oxygen during the time of flooding, which results in a diversity of root modifications (11.52-11.54). Generally, the root consumes the oxygen it has taken up through tap roots or through lenticels of root-knees at low tide. They consume oxygen at high tide and the reduced pressure then allows a refilling of the spongy aerenchyma again at low tide.

11.51 Growing in the tidal zone, besides salt, the roots of Mangroves must sustain periods of flooding without oxygen exchange. Hinchinbrook Island, Australia.

11.52 Aeration of roots also takes place at the tip of knee-like structures through thermo-osmosis of lenticels. Burma Mangrove (Bruguiera sp.), Blind-Your-Eye Mangrove (Excoecaria agallocha) and Apple Mangrove (Sonneratia alba). Hitchinbrook Island, Australia.

11.53 Root cross section of a Burma Mangrove (Bruguiera gymnorrhiza). A large proportion is occupied by the aerenchyma.

11.54 Similar knee-like structures are also formed by freshwater swamp plants, such as the Bald Cypress (Taxodium distichum). Okefenokee Swamp, Georgia, USA.

Other Special Ecological Adaptations
Mistletoes

Mistletoes are another functional group of woody species which have adapted to use other trees to settle and use resources of these hosts for their own survival. One may distinguish between phloem- and xylem-tapping mistletoes. The phloem-tapping mistletoes do not produce green leaves and are strictly parasitic (11.56). In contrast, the xylem-tapping mistletoes produce green leaves, and are parasitic mainly with respect to water and nutrients (11.57, 11.58). They also receive carbohydrates and amino acids skeletons as these substances are transported in the xylem. The attachment to the host is the critical step in the life cycle (11.55), because many of the mistletoes are species specific, i.e. they cannot attach to other hosts. See also p. 178.

11.55 After germination, the seed forms a haustorium instead of a root which dissolves the epidermis of the host twig and taps to the xylem. Broken Hill, Australia.

11.56 Phoradendron sp. on Sitka Spruce (Picea sitchensis) is an example of a phloem-tapping mistletoe. Mount Rainier, Washington, USA.

11.57 The Australian mistletoes are famous for the apparent similarity of leaves between host and mistletoes, which might originate from horizontal transfer of genes, and it is interpreted as mechanism against herbivory by marsupials.

11.58 Tapinanthus oleifolius of South Africa also parasitizes Euphorbia virosa. In this case, the mistletoe builds a long root-like haustorium which grows between the outer vascular tissue and the soft inner parenchyma.

Other Special Ecological Adaptations
Phyllodes, Phyllodes, Green Woody Stems

Mainly under arid conditions, but also in nutrient-limited habitats, some species reduce the leaf area in favor of a green petiole or rhachis (phyllodes; 11.59, 11.61), or flat green stems (phylloclades; 11.60) and round green woody stems (11.62) or roots (11.63). These may or may not utilise the Crassulacean Acid Metabolism as in succulents, and are also able to use the respiratory CO_2 of the stem.

11.59 Phyllodes of Black Wattle (Acacia decurrens, above) and Golden Wattle (Acacia pycnantha, below). Botanical Garden Bayreuth, Germany.

11.60 Phylloclades, leaf-like photosynthetic branches, of Mountain Toatoa (Phyllocladus alpina), Podocarpaceae. Botanical Garden Bochum, Germany (photo: Jagel).

11.61 Daviesia mollis, Papillionaceae. Here, the rhachis broadened to a leaf-like structure. The stem has also flattened to a phylloclade. Fitzgerald, Tasmania, Australia.

11.62 The formation of green stems is common under arid conditions, e.g. in Anabasis articulata, Cheniopodiaceae. The young bark is the assimilating organ. It is replaced by a secondary cork tissue in older parts. Negev, Israel.

11.63 Taeniophyllum filiforme is a leafless orchid with flat green aerial roots for assimilation. Botanical Garden Munich, Germany (photo: Gerlach).

CHAPTER 12

DECAY OF DEAD WOOD

This chapter illustrates anatomical features which are related to biological, physical and mechanical decay.

Deformed wood structure of an interglacial, compressed conifer branch.

Insects

Larvae and imagos of insects eat wood with their mandibles. This creates galleries just below the cambium, on the wood's surface and throughout the wood. Different insect species burrow different shapes of galleries (12.1-12.3). However, the anatomical structure is mostly unspecific (12.4-12.7). Since galleries in historical wood mainly destroyed the sapwood, and especially the most recent rings, it is often difficult to date these samples by dendrochronological means. □

12.1 Superficial insect galleries on a Two-needle Pinyon tree (Pinus edulis). Arizona, USA.

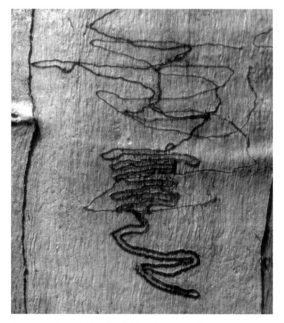

12.2 Marks of the Scribbly Gum Moth (Ogmograptis scribula). The larvae are feeding between the old and new season's bark of gum-barked Eucalypt trees. When the tree sheds the old bark the tunnels are revealed.

12.3 Hollow Eucalypt stem settled by termites.

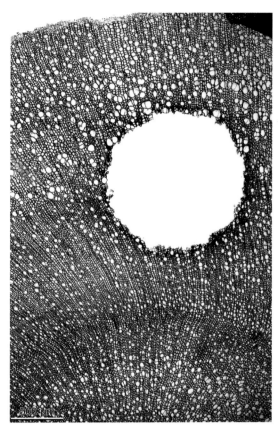

12.4 Insect gallery in a rhizome of the hemicryptic herb Hedge Bedtraw (Galium mollugo). The gallery is surrounded by a barrier zone.

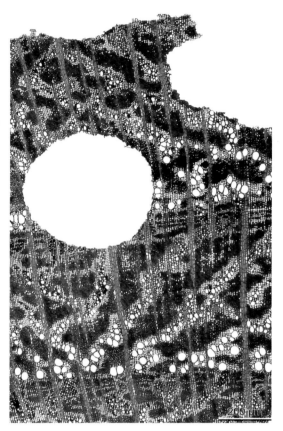

12.5 Insect galleries in Common Broom sapwood (Cytisus scoparius).

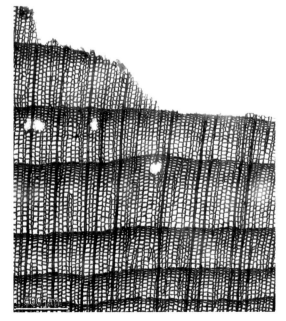

12.6 Superficial insect galleries on a Scots Pine (Pinus sylvestris). The former outermost ring is unevenly decomposed.

12.7 Thin section of a neolithic piece of fir charcoal with an insect gallery from the lake of Zürich, Switzerland. The piece had decayed due to fungi, before carbonization. Embedded in epoxy resin.

Fungi

Wood decomposition is mainly a result of fungal infestation (Schwarze et al. 1999; Schmidt 1994). Most wood decays under aerobic conditions. Advancing, spreading nets of hyphae create so-called demarcation lines in areas where they are highly concentrated (12.8). Thick, brown hyphae belong to the blue-stain group that discolours the wood (12.9). The thin hyphae are mostly species that decompose the wood (12.10): white rot fungi decay cellulose, hemicellulose and lignin. These species attack the cell walls like caries (12.11). Brown rot mainly consumes cellulose. Cell walls break and decay continuously (12.12). The hyphae of soft rot fungi burrow galleries into the cell walls and mainly decay the cellulose (12.13-12.16).

In the early stage of anaerobic bacterial decay, waterlogged wood develops mosaic-like patterns of normal and degraded cells; later on, all secondary cell walls disappear. The remaining wood only contains primary walls.

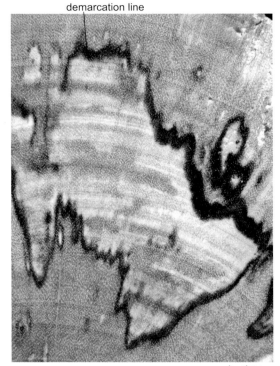

Right: 12.8 Demarcation lines in Silver Birch wood (Betula pendula). Note the irregular, dark lines around partially decomposed parts.

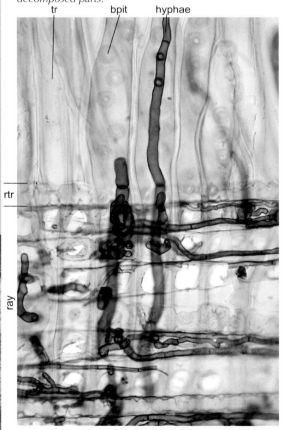

12.9 Thick, brown hyphae of a Blue-stain fungus in the tracheids and rays of a Scots Pine (Pinus sylvestris). Demarcation line.

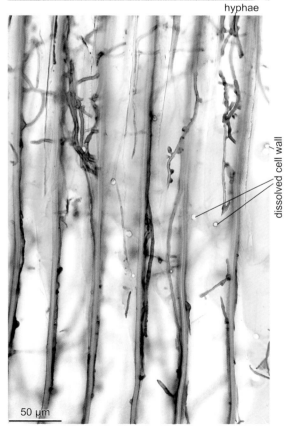

12.10 Thin, red-stained hyphae of a wood-decomposing fungus in Mountain Pine tracheids (Pinus mugo). They dissolve the cell walls enzymatically and create shoot-like holes. Brown rot in Pinus mugo.

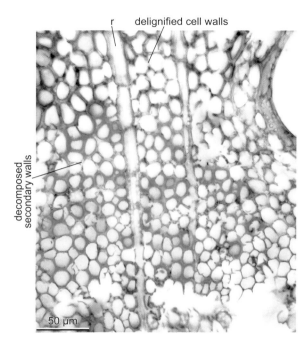

12.11 Symptoms of white rot fungi in Beech wood (Fagus sylvatica) with partially dissolved (light and dark red), often caries-like cell walls caused by this infestation.

12.12 Symptoms of white rot fungi in Heather wood (Calluna vulgaris) with blue, delignified borders of holes caused by this infestation.

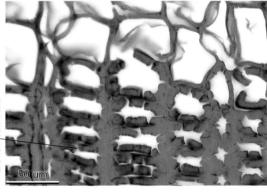

12.13 Symptoms of brown rot fungi in Mountain Pine latewood (Pinus mugo) with broken, unstructured cell walls caused by this fungus.

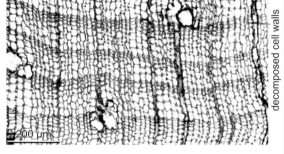

12.14 Symptoms of brown rot fungi. The cellulose is completely decomposed. There is no more reflection under polarized light. The absence of secondary walls is typical. The xylem consists of primary walls only. Due to decomposition, the density decreased from 0.6 to 0.2 g/cm³. The wood lost its stability. Mountain Pine (Pinus mugo).

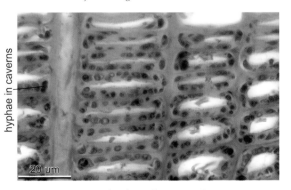

12.15 Symptoms of soft rot fungi. Hyphae create enzymatic caverns in the secondary walls of the latewood. Cross-section of Norway Spruce (Picea abies).

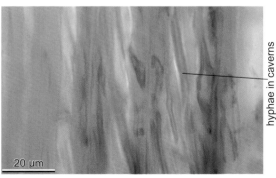

12.16 Symptoms of soft rot fungi. Hyphae follow the spiral-like, macrofibrillary latewood structure. Radial section of Norway Spruce (Picea abies).

12 DECAY OF DEAD WOOD

Carbonization

Carbonization is the result of a dry distillation process under anaerobic conditions. The residual charcoal mainly consists of carbon (12.17-12.19). Therefore, biological degradation is inhibited. During the process of carbonization, cell wall thickness is reduced; there is intensive weight loss and shrinkage (12.20). However, microscopic cell wall structures remain; hence it is still possible to determine the wood species (12.21; Schweingruber 1976).

12.17 One of the last charburners in Bavaria in 1969. Bayrischer Wald, Germany.

12.18 Charred remnants of a Scots Pine (Pinus sylvestris) after a severe forest fire. Bor Island, Siberia.

12.19 Even though the heating of the timber leads to the formation of charcoal, the effect is not very deep. A few millimeters below the surface the wood remains intact. It is decomposed by fungi unless another fire burns the dry timber.

12.20 Deliberately carbonized, dry Beech wood (Fagus sylvatica). The tree rings are perfectly visible. Only a few rays have become enlarged during the carbonization process (Schläpfer and Brown 1948).

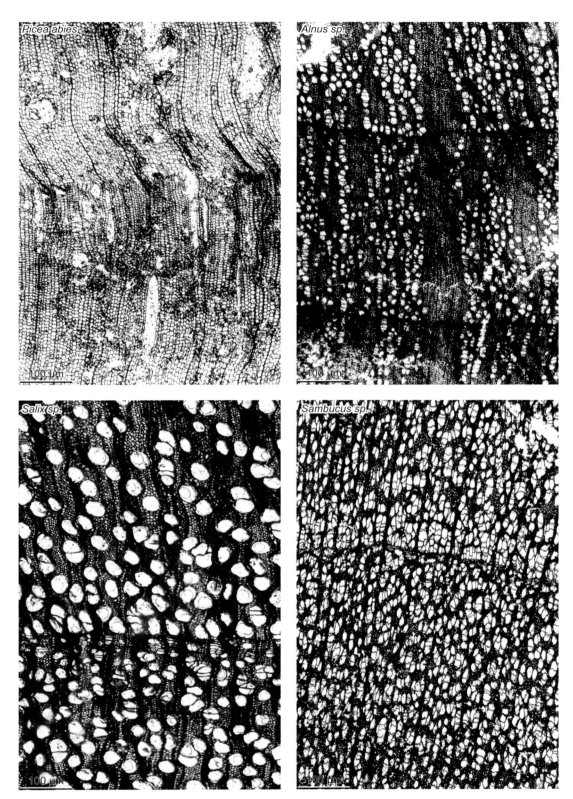

12.21 Thin section of charcoals. Characteristic is the dramatic shrinkage of cell walls through carbonization. Still, the principle anatomical structures survive the carbonization process. Presented here are samples from a Neolithic site in Switzerland. The samples were embedded in epoxy resin before cutting.

12 Decay of Dead Wood

Petrification

Petrified forests and tree stems (12.22, 12.23) are a palaeobotanical source of research material. In a complicated chemical process at low temperatures, all cavities in the wood are filled with minerals, in most cases with silicic acid, and the formerly organic material dissolves. The result is a perfect copy of the ancient cell wall structure (Dernbach *et al.* 1994). Hence, fossil wood can be identified (12.24–12.28).

12.22 Petrified forest in Arizona, USA.

12.23 Macroscopical aspect of a petrified Araucaria sp. stem. The growth zones are very distinct. Petrified forest in Arizona, USA.

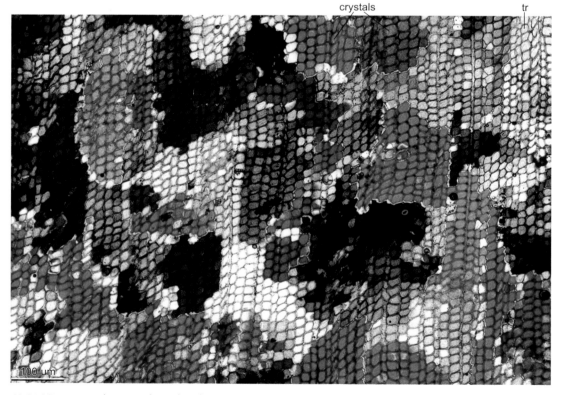

12.24 Microscopical aspect of a carboniferous Dadoxylon (Paradoxylon leuthardii) under polarized light. The colours reflect the crystalline quartz structure (slide: Buxtorf).

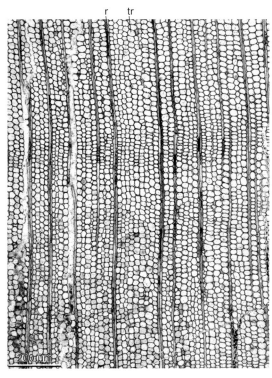

12.25 Cross-section of a carboniferous Dadoxylon. The zone with smaller tracheids indicates indistinct seasonal, climatic changes (slide: Buxtorf).

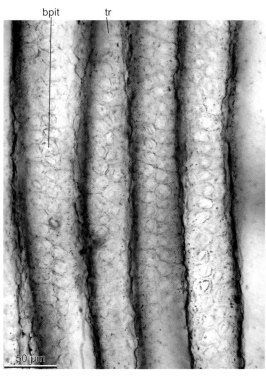

12.26 Radial section of a carboniferous Dadoxylon. Even microscopic cell wall characteristics are preserved in the silicified wood. Several rows of bordered pits are typical of the genus Dadoxylon (slide: Buxtorf).

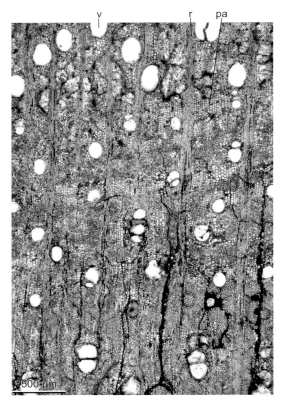

12.27 Petrified tertiary wood growing in a tropical climate. Note the absence of rings (slide: Buxtorf).

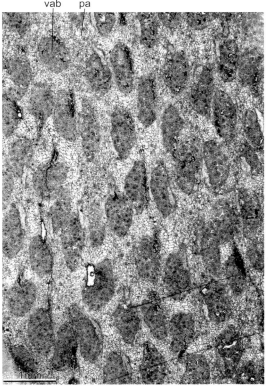

12.28 Petrified tertiary palm. The vascular bundles are characteristic of monocotyledonous plants (slide: Buxtorf).

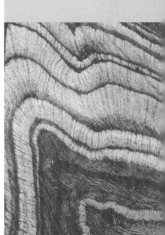

COMPRESSION

The anatomical structures of petrified and well-preserved fossil wood are often deformed (12.29, 12.30). If a heavy layer of sediment covers anaerobically decomposed, waterlogged wood, the cell wall structures become locally compressed (12.31, 12.32). Well-preserved conifer stems under an advancing glacier are also compressed (12.33, 12.34). In such cases, the light earlywood collapses, and only the dense latewood remains.

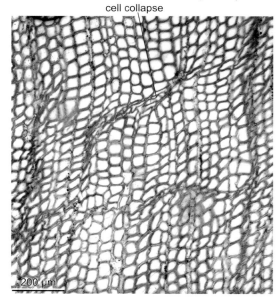

12.29 Collapse of a few tracheid rows in a petrified, carboniferous Dadoxylon stem. The stem was compressed by sediments (slide: Buxtorf).

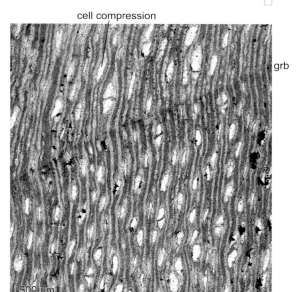

12.30 Deformed wood structure in the tertiary wood. The wood was compressed by sediments (slide: Buxtorf).

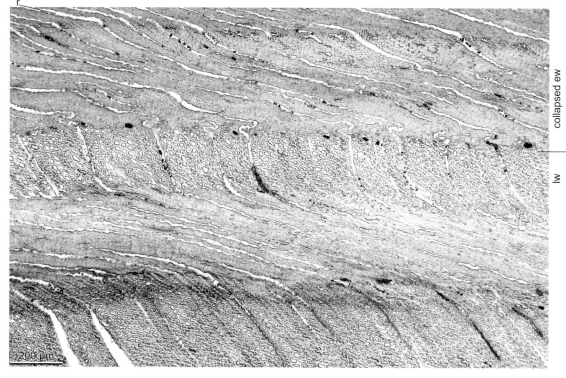

12.31 Deformed wood structure in the lateral part of an interglacial conifer branch. A heavy layer of sediment compressed and folded the thin-walled earlywood. On the basis of the bent rays, it is possible to estimate by how much the earlywood width was reduced. The latewood resisted the pressure (Picea sp. or Abies sp.; material: Welten).

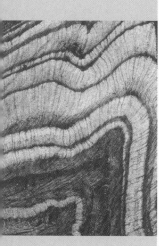

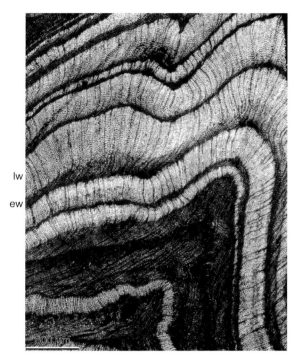
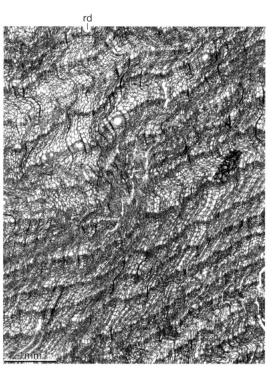

12.32 Deformed wood structure in the distal part of an interglacial, compressed conifer branch. Under polarized light, both the decomposition (dark lines) and the resistance of latewood cellulose (light bands) are clearly visible (Picea sp. or Abies sp.; material: Welten).

12.33 Deformed wood structure in a Holocene European Larch stem (Larix decidua). This was caused by the weight of a several-hundred-meter-thick layer of ice in the Great Aletsch Glacier in the Alps (material: Holzhauser).

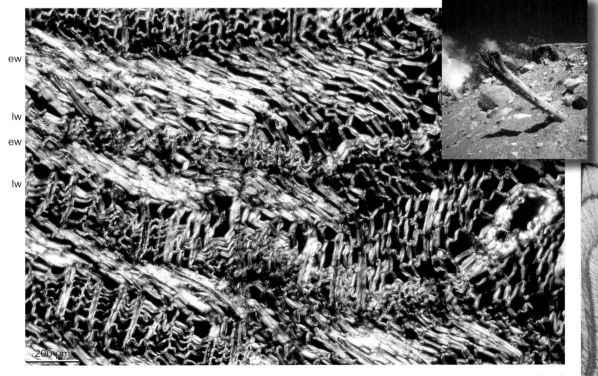

12.34 Collapsed earlywood and preserved latewood tracheids in a Larch stem, below the ice of the Great Aletsch Glacier in the Alps. The bent rays are typical (material: Holzhauser; polarized light). Small picture: A Larch stem in the moraine of a glacier (photo: Röthlisberger).

12 Decay of Dead Wood

Chapter 13

Microscopical Preparation

This chapter describes a few easily applicable wood anatomical techniques.

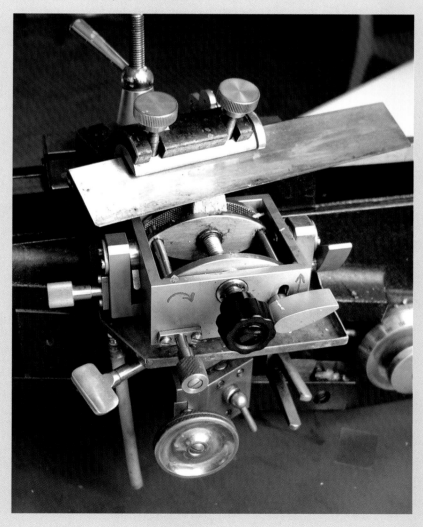

Sliding microtome. Wood sample and knife in correct position.

Collection and Storing of Material and Preparation for Sectioning

Fresh material is most suitable for microscopical preparation. For transportation and final storage, thick-walled plastic bags are best. Carefully washed plants, or plant parts, may be conserved for a long time in ethanol (40%) or any commercial alcohol, such as whisky or vodka, in thick, sealed plastic bags.

The plant samples are labelled with a very soft pencil, e.g. Stabilo-Aquarellable, on thick paper. It is important to record the following details: Latin name, plant part (rhizome, primary root, shoot, transition etc.), life form, phenological plant stage, stem deformation (if any), site conditions e.g. dry slope, beech forest, grazed etc., locality, region, country, altitude, collecting date.

13.1 Wood samples mounted on grafting wax for episcopic observation.

Dry wood is suitable for anatomical studies of wood, but the cell contents, for example the nuclei, are denatured and thin walled tissues in the bark have collapsed. Soft wood must be moistened, and very hard wood cooked in water for several minutes or hours before sectioning. Very hard wood softens after having been soaked in an equal mixture of water, ethanol (96%) and glycerol for some weeks.

Sample preparation depends on the size of the plants. For episcopic observations and microscopical preparation, about 1 cm³ pieces are cut from big or whole stems (13.1). Before sectioning, radial and tangential directions have to be splitted with a knife. Very thin stems are clamped in polystyrene and can be stored in upright position in Petri dishes.

Making thin Sections

If one has a very steady hand, it is possible to cut thin sections with a razor blade. In order to obtain high quality sections, however, a sliding microtome is indispensable. Sliding microtomes, e.g. by REICHERT or EUROMEX, are best (13.2). A pre-requisite for good sections is a perfectly sharpened knife (Type C). Sharpening by hand is possible with a cheap sharpening tool by EUROMEX. Very sharp paper knife blades (NT) can be clamped in knife holders (product of EUROMEX; 13.3). Disposable blades may be used for small soft pieces of wood and herbaceous plants.

By following a few rules, even beginners can produce good sections: trickle absolute alcohol on the plane surface of soft materials, press an aquarelle brush on the sample surface and pull the knife at an inclination of approximately 15°, the disposal blade at 3° in a very steep angle (13.2) below the brush. At least a 5 cm long portion of the knife is needed for each section. This procedure prevents the section from rolling up. Pull the section in much alcohol with the brush on the slide. The thickness of the section varies between 10 and 60 µm. Even on thick sections, ring boundaries may be clearly recognized.

Sections may be kept on microscope slides for hours or days, after a little bit of glycerol has been dropped onto them.

13.2 An old REICHERT-sliding microtome. The perfect orientation of the section may be achieved with the sample holder, which can be moved three-dimensionally. A perfectly sharpened knife slides into the V-shaped groove. The best cuts are obtained by using a blade at least 5 cm long, as shown in the photograph.

13.3 Blade holder. Special blades for microtomy or paper knife blades (NT or other products) are suitable for cutting soft herb stems and soft, watersaturated wood.

Preparation of Thin Sections for Permanent Slides

All preparatory steps are carried out on the slide. The solutions are trickled directly onto the section, and allowed to run off it again by holding the slide askew. The waste is collected in a glass.

There are many staining methods (Gerlach 1984; Chaffey 2002). Normally, astrablue (Astrablue 0.5 g, or acetic or tartaric acid 2 ml, distilled water 100 ml) and safranin (Safranin 1% water solution) are used. Both colours are mixed in equal proportions, and the dye is trickled onto the section for 3-5 min. Afterwards, the sample gets dehydrated (water, 95% alcohol, xylol, Canada Balsam).

Dehydration is achieved with alcohol 95% and then with a mixture of 95% alcohol and 5% of 2,2-dimethoxypropane-acetone-dimethylacetate (FLUKA). The alcohol is dripped several times onto the section on the slide, as described above. When no more dye flows out of the section, xylol is trickled onto it. Dehydration is complete when the xylol does not turn milky. Finally, a little drop of Canada Balsam is put on the section, which is closed with a cover glass. In our experience, Canada Balsam is the best and most permanent embedding resin. Many sections remain uneven, making observation difficult.

13.4 Slides between thick plastic strips on an iron plate. The magnets press the section flat.

Therefore, the microscope slides are compressed with a little magnet between two plastic strips on an iron plate, whilst the section is drying out in the oven (60 °C for 12 h; 13.4). The hard resin on and outside the cover glass may be scraped off with a razor blade. Never use xylol.

The anatomy of species containing slime, phenolic substances or much starch is difficult to observe. In those cases, at first, Eau de Javelle (potassiumhypochlorite, KOCl) is trickled onto the section for 5-10 min. All cell contents, e.g. nuclei and starch, are destroyed. Afterwards, the section is rinsed with water until the odour of the Eau de Javelle has disappeared. Then we can stain and dehydrate the sample.

Observation and Photography

Generally, normal transmission light is used. Unlignified cell walls appear blue, lignified ones red. By using dark-green filters, the contrast of pale tissue may be increased. Naturally dark material may be observed without staining. Polarized light is extremely useful for examining the stages of cell wall development and degradation in secondary vascular tissue. One polarization filter is placed below the eye-piece, and another on the light source below the objective. By turning one filter, cells with secondary walls appear light, and cell wall structures without a spiral-shaped, macrofibrillar structure disappear (13.5). For photography we use microscopes with digital camera equipment (13.6).

 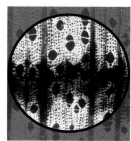

13.5 Demarcation line in an unstained cross section of Sycamore (Acer pseudoplatanus). Normal light: light background (left), and polarized light: dark background (right).

13.6 Slide under the microscope with mounted digital camera.

Addicot FT (1981) *Abscission*. Berkley, Univ. of California Press

Aeschimann D, Lauber K, Moser DM, Theurillat J-P (2004) *Flora alpina*. Haupt, Bern, Stuttgart, Wien. 3 Vol

Baas P, Schweingruber FH (1987) *Ecological trends in the wood anatomy of trees, shrubs and climbers in Europe*. IAWA Bull 8:245-287

Bailey IW (1923) *The cambium and its derivate tissues*. Am J Bot 10:499-509

Baltensweiler W, Rubli D (1999) *Dispersal: an important driving force of the cyclic population dynamics of the larch bud moth, Zeiraphera diniana Gn*, pp 3-153

Bormann FH (1962) *Root grafting and non-competitive relationships between trees*. In: Kozlowski TT (ed) *Tree growth*. Ronald Press Company, New York

Bosshard HH (1982) *Holzkunde*. Vol.2. Birkhäuser, Basel, Boston, Stuttgart

Briffa KR, Osborn TJ, Schweingruber FH, Jones PD, Shiyatov SG, Vaganov EA (2002) *Tree-ring width and density around the Northern Hemisphere: Part 2, spatio-temporal variability and associated climate patterns*. The Holocene 12:759-789

Carlquist S, Schneider EL (2001) *Vegetative anatomy of the New Caledonian endemic Amborella trichopoda: Relationships with the Illiciales and implications for vessel origin*. Pacific Sci 55:305-312

Chaffey N (2002) *Wood formation in trees. Cell and molecular biology techniques. Chapter 2: Wood microscopical techniques*. Taylor and Francis, London and New York, pp 17-40

Cherubini P, Schweingruber FH, Forster T (1997) *Morphology and ecological significance of intra-annual radial cracks in living conifers*. Trees 11:216-222

Christensen K (1987) *Tree-rings and insects: The influence of cockchafers on the development of growth rings in oak trees*. In: Jacoby GC Jr., Hornbeck JW (Compilers) *Proceedings of the International Symposium on Ecological Aspects of Tree-Ring Analysis*, August 17-21, 1986, Tarrytown, New York. U.S. Department of Energy, Publication CONF-8608144: 142-154

Core HA, Coté WA, Day AC (1979) *Wood structure and identification*. 3rd edn. Syracuse Univ. Press.

Dernbach U, Glas M, Hochleitner R, Jung W, Landmesser M, Mayr H, Selmeier A (1994) *Versteinertes Holz*. Extra Lapis 7. Christian Weise, München

Duff HG, Nolan NJ (1953) *The controls of cambial and apical activity of Pinus resinosa Ait*. Can J Bot 31:471-513

Endress P, Igersheim A (2000) *The reproductive structures of the basal Angiosperm Amborella trichopoda (Amborellaceae)*. Int J Plant Sci 16 (6 Suppl.):237-248

Esau K (1977) *Anatomy of seed plants*. Wiley, New York

Esper J, Schweingruber FH (2004) *Large-scale treeline changes recorded in Siberia*. Geophysical Research Letters 31, L06202, doi:10.1029/2003GL019178

Filion L, Payette S, Gauthier L, Boutin Y (1986) *Light rings in subarctic conifers as a dendrochronological tool*. Quat Res 26:272-279

Fischer U, Heinzler M, Kilgus R, Näher F, Osterle S, Paetzold H, Röhrer W, Stephan A, Winkow R (2002) *Tabellenbuch Metall*, 42. Aufl. Europa-Lehrmittel, Haan-Gruiten

Gerlach D (1984) *Botanische Mikrotechnik*, 3rd edn. Thieme, Stuttgart, New York

Goldammer G, Furyaev VV (eds) (1996) *Fire in ecosystems of Boreal forests*. Forestry sciences 48. Kluwer Acacemic Press

Henes E (1959) *Fossile Wandstrukturen untersucht am Beispiel der Tracheidenwände paläozoischer Gefäßpflanzen*. Encyclopedia of plant anatomy. Gebrüder Bornträger, Berlin-Nikolassee

Höster HR, Liese W (1966) *Über das Vorkommen von Reaktionsgewebe in Wurzeln und Ästen der Dicotyledonen*. Holzforschung 20:80-90

Höster HR, Liese W, Böttcher P (1968) *Untersuchungen zur Morphologie und Histologie der Zweigabwürfe von Populus „Robusta"*. Forstwiss. Zentralblatt 87:365-368

Huber B (1961) *Grundzüge der Pflanzenanatomie*. Springer, Berlin Göttingen Heidelberg

Jalkanen R, Alto T and Kurkela T (1995) *Development of needle retention in Scots Pine (Pinus sylvestris) in 1957-1991 in northern and southern Finland*. Trees, 10: 125-133

Judd W, Campbell CS, Kellog EA, Stevens PF, Donoghue MJ (2002) *Plant systematics*. Sinauer Assoc. Inc. Sutherland, Massachusetts, USA

Jung M, Henkel K, Herold M, Churkina G (2006) *Exploiting synergies of global land cover products for carbon cycle modeling*. Remote Sensing of Environment 101: 534-553

Jung W (1986) *Die Idee „Baum"*. Aktuelle Geoinformation. Sonderschau der paläontologischen Staatssammlung München. 17-19 October 1986

Körner C (1999) *Alpine plant life*. Springer, Berlin Göttingen Heidelberg

Kräusel R, Weyland H (1935) *Neue Pflanzenfunde im rheinischen Unterdevon*. Palaeontographica 80, B:171-190

Larcher W (2001) *Ökophysiologie der Pflanzen*. 6. Aufl., Ulmer, Stuttgart

Mägdefrau K (1951) *Botanik. Eine Einführung in das Studium der Pflanzenkunde*. Winter, Heidelberg

Mattheck C, Kübler H (1995) *Wood-internal optimization of trees*. Springer Series in Wood Sciences. Springer, Berlin Heidelberg New York

Meusel H (1976) *Die Evolution der Pflanzen in pflanzengeographisch-ökologischer Sicht*. In: Böhme H, Hagemann R, Löther R (eds) *Beiträge zur Abstammungslehre*. Berlin

Musaev MK (1996) *Seasonal growth and anatomic structure of annual rings in Pinus sylvestris L. in the area of the Chernobyl catastrophe*. Lesovedenie 0(1):16-29

Pierozynski KS, Malloch DW (1975) *The origin of land plants: a matter of mycotropism*. Bio Systems 6:153-164.

Roloff A (2001) *Baumkronen. Verständnis und praktische Bedeutung eines komplexen Naturphänomens*. Ulmer, Stuttgart

Scharschmidt F (1968) *Paläobotanik I*. Hochschultaschenbücher, Bibliographisches Institut. 357/557a

Schläpfer P, Brown R (1948) *Über die Struktur der Holzkohlen*. Eidgenössische Materialprüfungs- und Versuchsanstalt für Industrie, Bauwesen und Gewerbe, Zürich. Bericht 153

Schmidt O (1994) *Holz und Baumpilze: Biologie, Schäden, Schutz, Nutzen*. Springer, Berlin Göttingen Heidelberg

Schulze ED, Beck E, Müller-Hohenstein K (2005) *Plant Ecology*. Springer, Berlin Heidelberg New York

Schwarze FW, Engels J, Mattheck C (1999) *Holzzerstörende Pilze in Bäumen*. Rombach-Verlag, Freiburg im Breisgau

Schweingruber FH (1976) *Prähistorisches Holz. Die Bedeutung von Holzfunden aus Mitteleuropa für die Lösung archäologischer und vegetationskundlicher Probleme*. Academia Helvetica 2. Haupt, Bern Stuttgart Wien

Schweingruber FH (1978) *Mikroskopische Holzanatomie*. 3. Aufl. Eidg. Forschungsanstalt für Wald Schnee und Landschaft

Schweingruber FH (1996) *Tree rings and environment, dendroecology*. Haupt, Bern Stuttgart Wien

Schweingruber FH (2001) *Dendroökologische Holzanatomie. Anatomische Grundlagen der Dendrochronologie*. Haupt, Bern Stuttgart Wien

Schweingruber FH, Poschlod P (2005) *Growth rings in herbs and shrubs; life span, age determination and stem anatomy*. Forest Snow and Landscape Research 79: 195–415

Schweizer HJ (1990) *Pflanzen erobern das Land*. Kleine Senckenberg-Reihe 18:1-74

Scotese CR (1997) *Paleogeographic Atlas*. PALEOMAP Progress Report 90-0497, Department of Geology, University of Texas at Arlington, Arlington, Texas, 37 pp

Sell J (1987) *Eigenschaften und Kenngrössen von Holzarten*. Baufachverlag AG Zürich Dietikon

Shigo AL (1989) *A new tree biology. Facts, photos and philosophies in trees and their problems and proper care*. 2nd edn. Shigo and Trees Ass., Durham, New Hampshire

Skomarkova MV, Vaganov EA, Mund M, Knohl A, Linke P, Börner A, Schulze ED (in prep.) *Inter-annual and seasonal variability of radial growth, wood density and carbon isotope ratios in tree rings of beech (Fagus sylvatica) growing in Germany and Italy*

Strasburger E, Noll F, Schenk H, Schimper AFW (2002) *Lehrbuch der Botanik*. 35. Auflage. Neubearbeitet von Sitte P, Weiler EW, Kaderit IW, Bresinsky A, Körner C. Spektrum Akademischer, Heidelberg Berlin

Swetnam T, Lynch AM (1993) *Multicentury, regional-scale patterns of Western spruce budworm outbreaks*. Ecol Monographs 63:399-424

Timell TE (1986) *Compression wood in Gymnosperms*. 3 Vol. Springer, Berlin Heidelberg New York

Vaucher H (1990) *Baumrinden*. Enke-Verlag, Stuttgart

Wagenführ R (1989) *Anatomie des Holzes*. 4. Aufl. VEB Fachbuchverlag Leipzig

Walter H (1968) *Die Vegetation der Erde, II. Die gemäßigten und arktischen Zonen*. Fischer, Jena Stuttgart

Wink M (2005) *Schriftzeichen im Logbuch des Lebens: Molekulare Evolutionsforschung*. Biol. Unserer Zeit 1/2006 (36):26-37

Worbes M (1994) *Grundlagen und Anwendung der Jahrringforschung in den Tropen*. Habilitationsschrift, Univ. Hamburg

Zhou Z, Barett PM, Hilton J (2003) *An exceptionally preserved Lower Cretaceous ecosystem*. Nature 421:807-814

Zimmermann W (1959) *Die Phylogenie der Pflanzen*. Fischer, Stuttgart

Scientific Name	Family	Common Name	Page
Abies alba	Pinaceae	Silver Fir	23, 76, 86, 170
Abies pinsapo	Pinaceae	Spanish Fir	100
Acacia abyssinica	Fabaceae	Flat-top Acacia	80, 144, 148
Acacia cymbispina	Fabaceae		199
Acacia decurrens	Fabaceae	Black Wattle	202
Acacia drepanolobium	Fabaceae	Flute Acacia	199
Acacia erubescens	Fabaceae	Blue Thorn	28
Acacia hotwittii	Fabaceae		177
Acacia mellifera	Fabaceae	Black-thorn Acacia	195
Acacia pycnantha	Fabaceae	Golden Wattle	200
Acacia reficiens	Fabaceae	Red Umbrella Acacia	195
Acer campestre	Acearaceae	Common Maple	192
Acer pseudoplatanus	Acearaceae	Sycamore	66, 103, 149, 159, 178
Adansonia digitata	Bombaceae	African Baobab	198
Adansonia gregorii	Bombaceae	Australian Baobab	198
Adenocarpus villosus	Fabaceae	Codeso de Cumbre	37
Aeonium arboreum	Crassulaceae	Black Tree Aeonium	44
Aesculus hippocastanum	Hippocastanaceae	Horse Chestnut	169, 170
Agathis australis	Araucariaceae	Kauri	190
Alnus glutinosa	Betulaceae	Black Alder	115
Alnus incana	Betulaceae	Grey Alder	75, 127, 160, 161
Alnus viridis	Betulaceae	Green Alder	180
Aloe dichotoma	Aloaceae	Kokerboom	194
Amborella trichopoda	Amborellaceae	Amborella	24
Anabasis articulata	Brassicaceae		202
Anisum sativum	Apiaceae	Anise	103
Araucaria angustifolia	Araucariaceae	Parana Pine	23
Araucaria araucana	Araucariaceae	Monkey-puzzle Tree	22, 100
Archaeopteris	Archaeopteridales		16
Arctostaphylos uva-ursi	Ericaceae	Red Bearberry	61, 62, 158
Argania spinosa	Sapotaceae	Argan Tree	166
Aristolochia clematitis	Aristolochiacea	Birthwort	25
Asparagus tenuifolium	Liliaceae	Asparagus	41
Asplenium viride	Aspleniaceae	Green Spleenwort	15
Asteroxylon mackiei	Asteroxylaceae		44
Atriplex sp.	Chenopodiaceae	Saltbush	71
Azorella compacta	Apiaceae	Yareta	196
Banksia brownii	Proteaceae	Brown's Banksia	29
Banksia serrata	Proteaceae	Saw Banksia	92
Berberis sp.	Berberidaceae	Barberry	123, 169
Berteroa incana	Brassicaceae	Hoary Alison	56
Betula nana	Betulaceae	Dwarf Birch	147
Betula pendula	Betulaceae	Silver Birch	55, 119, 144, 148, 164, 165, 180, 206
Betula pubescens	Betulaceae	Downy Birch	29
Bougainvillea spectabilis	Nyctaginaceae	Bougainvillea	27, 91
Brachychiton rupestre	Sterculiaceae	Queensland Bottle Tree	198
Bruguiera gymnorrhiza	Combretaceae	Burma Mangrove	200
Bryonia dioica	Cucurbitaceae	White Bryony	93
Buxus sempervirens	Buxaceae	Box Wood	35, 169
Calamites sp.	Calamithaceae		11
Calluna vulgaris	Ericaceae	Heather	150, 204
Capparis sp.	Capparidaceae	Caper	91
Capsella bursa-pastoris	Brassicaceae	Shepherd's Purse	139
Carex pendula	Cyperaceae	Pendulous Sedge	71
Carica papaya	Caricaceae	Papay	65
Carlina sp.	Asteraceae	Carline Thistle	193
Carnegiea gigantea	Cactaceae	Saguaro	194
Carpinus betulus	Betulaceae	Common Hornbeam	74, 103
Castanea sativa	Fagaceae	Sweet Chestnut	58, 59, 61, 79, 113, 149, 150, 151, 192, 157, 172, 175, 180
Ceiba pentandra	Bombacaceae	Kapok Tree	186
Chamaedaphne calyculata	Ericaceae	Leather-leaf	147

List of Species

Scientific Name	Family	Common Name	Page
Chenopodium album	Chenopodiaceae	White Goosefoot	91
Chenopodium sp.	Chenopodiaceae	Goosefoot	91
Chosenia arbutifolia	Salicaceae		62
Cirsium arvense	Asteraceae	Creeping Thistle	88
Clematis vitalba	Ranunculaceae	Traveller's Joy	74
Cocos nucifera	Arecaceae	Coconut Palm	82
Commiphora saxicola	Burseraceae	Rock Corkwood	198
Cooksonia devonica	Rhyniaceae		7
Corema alba	Empetraceae		128
Corylus avellana	Betulaceae	Hazelnut	73, 113, 128, 129, 169
Crataegus monogyna	Rosaceae	One-seed Hawthorn	118
Cyathea capensis	Cyatheaceae		64
Cycas armstrongii	Cycadacea		18, 19
Cytisus scoparius	Fabaceae	Common Broom	205
Dadoxylon sp.	Cordaitaceae		17, 212
Daphne mezereum	Thymelaeaceae	Mezereon	89
Daphne striata	Thymelaeaceae	Garland Flower	73
Daviesia mollis	Fabaceae		202
Dicksonia antarctica	Dicksoniaceae	Tasmanian Treefern	14
Digitalis obscura	Scrophulariaceae	Dusky Foxglove	108
Dracaena draco	Dracenaceae	Dragon Tree	83, 85
Dracaena serrulata	Dracenaceae	Dragon Tree	85
Dryas integrifolia	Rosaceae	Arctic Mountain Avens	28
Drymis piperita	Winteraceae		144
Dryopteris austriaca	Aspidiaceae	Broad Buckler Fern	15
Dryopteris filix-mas	Aspidiaceae	Male Fern	15
Ephedra helvetica	Ephedraceae	Joint Pine	81
Ephedra major	Ephedraceae	Joint Pine	21
Ephedra sp.	Ephedraceae	Joint Pine	21
Equisetum hiemale	Equisetaceae	Rough Horsetail	11
Equisetum maximum	Equisetaceae	Giant Horsetail	11
Eremophila sp.	Myoporaceae	Emu Bush	177
Eucalyptus diversicolor	Myrtaceae	Karri	31
Eucalyptus jacksonii	Myrtaceae	Red Tingle Tree	191
Eucalyptus marginata	Myrtaceae	Jarrah	190
Eucalyptus pauciflora	Myrtaceae	Snow Gum	108, 176
Eucalyptus salmonophloia	Myrtaceae	Salmon Gum	30
Eucalyptus socialis	Myrtaceae	Mallee Eucalypt	198
Eucalyptus sp.	Myrtaceae	Eucalypt	30, 150, 174
Euphorbia cyparissias	Euphorbiaceae	Cypress Spurge	138
Euphorbia pescatoria	Euphorbiaceae		145
Euphorbia virosa	Euphorbiaceae		201
Excoecaria agallocha	Euphorbiaceae	Blind-Your-Eye Mangrove	200
Fagus sylvatica	Fagaceae	European Beech	20, 33, 36, 48, 57, 79, 80, 98, 99, 105, 119, 129, 157, 166, 170, 171, 172, 173, 185, 207, 208
Ficus carica	Moraceae	Common Fig	109, 127
Ficus obliqua	Moraceae	Strangler Fig	197
Filipendula ulmaria	Rosaceae	Meadowsweet	71, 142
Fraxinus excelsior	Oleaceae	Common Ash	40, 49, 63, 103, 133, 136, 154, 161, 166, 167, 168, 185
Galium mollugo	Rubiaceae	Hedge Bedtraw	205
Ginkgo biloba	Ginkgoaceae	Ginkgo	18, 19
Gleditsia triacantho	Fabaceaes	Honey Locust	122, 169
Globularia cordifolia	Globulariaceae	Heart-leaved Daisy	137
Gnetum gnemon	Gnetaceae		20
Guijera parviflora	Rutaceae		145
Hakea sp.	Proteaceae		78
Haloxylon persicum	Chenopodiaceae	Salt Tree	108
Hedera helix	Araliaceae	Ivy	36, 101, 172
Helleborus viridis	Ranunculaceae	Green Hellebore	103, 132

Scientific Name	Family	Common Name	Page
Hippocrepis comosa	Fabaceae	Horseshoe Vetch	128
Hippophae rhamnoide	Eleagnaceaes	Sea Buckthorn	49, 79, 151, 158
Hordeum vulgare	Gramineae	Barley	73
Hyenia elegans	Hyeniales		8
Hypericum coris	Hypericaceae	Heath-leaved St. Johns-worth	88
Ilex aquifolium	Aquifoliaceae	Common Holly	101, 126, 146, 151
Juglans nigra	Juglandaceae	Black Walnut	51
Juglans regia	Juglandaceae	Common Walnut	34, 35, 36, 40, 66, 151, 181
Juncus arcticus	Juncaceae	Arctic Rush	81
Juniperus communis	Cupressaceae	Common Juniper	47
Juniperus depeana	Cupressaceae	Alligator Juniper	23
Juniperus nana	Cupressaceae	Dwarf Juniper	190
Juniperus sabina	Cupressaceae	Savin Juniper	37, 51, 87
Juniperus sp.	Cupressaceae	Juniper	115
Juniperus turkestanica	Cupressaceae		22
Laburnum anagyroides	Fabaceae	Golden Rain	89, 152
Larix decidua	Pinaceae	European Larch	52, 102, 120, 143, 162
Larix gmelinii	Pinaceae	Dahurian Larch	54, 65, 113
Larix laricina	Pinaceae	Tamarack	153, 159
Larix sibirica	Pinaceae	Siberian Larch	110, 158, 159, 164, 165
Lebachia piniformis	Voltziales		22
Lemaireocereus weberi	Cactaceae	Cardon	194
Lepidodendron sp.	Lepidodendraceae		10
Leucanthemum vulgare	Asteraceae	Margherita	127
Limonium caprariense	Plumbaginaceae		196
Linum usitatissimum	Linaceae	Linseed	32
Livistonia sp.	Arecaceae	Cabbage Palm	82
Lobelia telekii	Campanulaceae	Giant Lobelia	195
Loiseleuria procumbens	Ericaceae	Mountain Azalea	35, 149
Lonicera pyrenaica	Caprifoliaceae	Pyrenean Honeysuckle	81
Lycopodium alpinum	Lycopodiaceae	Alpine Clubmoss	72
Lythrum salicaria	Lythraceae	Purple Loosestrife	53, 79
Macrozamia communis	Zamiaceae		18, 19
Magnolia stellata	Magnoliaceae	Star Magnolia	25
Magnolia virginianum	Magnoliaceae		25
Malcolmia aegyptica	Brassicaceae		148
Malus sylvestris	Rosaceae	Apple Tree	104, 114, 122
Malva parviflora	Malvaceae	Least Mallow	71
Melia azedarach	Meliaceae	Chinaberry	78
Mentha longifolia	Lamiaceae	Horse Mint	63
Mespilus germanica	Rosaceae	Medlar	33, 48
Microberlinia sp.	Fabaceae	Zebrawood	44
Mulinum spinosum	Apiaceae		28
Myricaria germanica	Tamaricaceae	German Tamarisk	88
Myrmecodia sp.	Rubiaceae		199
Neoraimondia arequipensi	Cactaceae		194
Nepenthes sp.	Nepenthaceae		197
Nolana sp.	Solanaceae	Nolana	145
Nothofagus pumilio	Fagaceae		110, 159
Nuytsia floribunda	Loranthaceae	Christmas Tree	90
Olea europaea	Oleaceae	Olive Tree	133
Ononis spinosa	Fabaceae	Spiny Restharrow	128
Opuntia sp.	Cactaceae	Prickly Pear	142
Osmunda regalis	Osmundaceae	Royal Fern	13
Ostrya carpinifolia	Betulaceae	Hop Hornbeam	89
Pachypodium namaquensis	Apocyanaceae		196
Paeonia suffruticosa	Paeoniaceae	Mountain Peony	80
Paradoxylon leuthardii	Cordaitaceae	Dadoxylon	210

List of Species

Scientific Name	Family	Common Name	Page
Phoenix canariensis	Arecaceae	Canarian Date Palm	31
Phoenix dactylifera	Arecaceae	Date Palm	30
Phoradendron sp.	Loranthaceae		201
Phyllocladus alpinus	Podocarpaceae	Mountain Toatoa	202
Phyllostachys edulis	Gramineae	Bamboo	73
Picconia azorica	Lauraceae	Pau-branco	133
Picea abies	Pinaceae	Norway Spruce	30, 36, 45, 47, 50, 52, 54, 57-59, 73, 76, 77, 87, 92, 93, 102, 124, 125, 128, 129, 141, 154, 161, 170, 171, 172, 180, 182, 193, 207
Picea engelmannii	Pinaceae	Engelmann Spruce	163, 164
Picea mariana	Pinaceae	Black Spruce	22, 28, 188
Picea obovata	Pinaceae	Siberian Spruce	164
Picea sitchensis	Pinaceae	Sitka Spruce	201
Pinus canariensis	Pinaceae	Canary Pine	64, 175
Pinus cembra	Pinaceae	Stone Pine	39, 43, 58, 61, 93, 104, 125, 156
Pinus edulis	Pinaceae	Two-needle Pinyon	204
Pinus leiophylla	Pinaceae	Smooth-leaved Pine	160
Pinus leucodermis	Pinaceae	Bosnian Pine	166
Pinus longaeva	Pinaceae	Bristlecone Pine	189, 196
Pinus mugo	Pinaceae	Mountain Pine	53, 58, 98, 110, 113, 121, 180, 182, 184, 185, 206, 207
Pinus nigra	Pinaceae	European Black Pine	71
Pinus pinaster	Pinaceae	Maritime Pine	150
Pinus pinea	Pinaceae	Umbrella Pine	151
Pinus ponderosa	Pinaceae	Ponderosa Pine	176, 177
Pinus radiata	Pinaceae	Monterey Pine	175
Pinus sylvestris	Pinaceae	Scots Pine	23, 31, 35, 60, 76, 108, 112, 155, 165, 176, 177, 178, 182, 185, 205, 206, 208
Platanus orientalis	Platanaceae	Oriental Plane	65
Podocarpus falcatus	Podocarpaceae	Outeniqua Yellowwood	48
Polytrichum commune	Polytrichales	Hair Cap Moss	69, 70, 72
Populus nigra	Salicaceae	Black Poplar	79
Populus sp.	Salicaceae	Cottonwood	59, 79, 99, 185
Potentilla fruticosa	Rosaceae	Shrubby Cinquefoil	135
Potentilla micrantha	Rosaceae	Barren Strawberry	135
Prasium majus	Lamiaceae	Hedge nettle	133
Prosopis glandulosa	Fabaceae	Honey Mesquite	104
Prunus avium	Rosaceae	Wild Cherry	134, 146, 155
Prunus domestica ssp. *syriaca*	Rosaceae	Syrian Plum	188
Prunus fruticosa	Rosaceae	Ground Cherry	134
Prunus mahaleb	Rosaceae	St. Lucie's Cherry	111
Prunus persica	Rosaceae	Peach Tree	99
Prunus spinosa	Rosaceae	Blackthorn	122
Psaronius	Marattiales		12
Psaronius brasiliensis	Marattiales		13
Pseudobombax mungaba	Bombacaceae		144
Pseudotsuga menziesii	Pinaceae	Douglas Fir	77, 163
Psilotum nudum	Psilotaceae	Whisk Fern	9, 72
Pteridium aquilinum	Denstaediaceae	Bracken Fern	15, 72
Pterocarpus angolensis	Fabaceae	African Teak	148
Pulsatilla vernalis	Ranunculaceae	Spring Anemone	132
Punica granatum	Punicaceae	Punica	109
Pyrus communis	Rosaceae	Pear Tree	128
Quercus coccifera	Fagaceae	Kermes Oak	150
Quercus petraea	Fagaceae	Sessile Oak	97, 109, 163, 179, 184
Quercus pubescens	Fagaceae	Pubescent Oak	80, 109, 129, 152, 160, 177
Quercus robur	Fagaceae	Common Oak	56, 75, 78, 159, 163, 181
Quercus sp.	Fagaceae	Oak	45
Quercus suber	Fagaceae	Cork Oak	88, 92, 172
Ranunculus acer	Ranunculaceae	Meadow Buttercup	132
Raoulia tenuicaulis	Asteraceae	Mat Daisy	128
Rhamnus alpina	Rhamnaceae	Alpine Buckthorn	134
Rhamnus cathartica	Rhamnaceae	Common Buckthorn	192
Rhamnus pumila	Rhamnaceae	Dwarf Buckthorn	134

Scientific Name	Family	Common Name	Page
Rhododendron ferrugineum	Ericaceae	Rusty-leaved Alpenrose	89, 111
Rhododendron parviflora	Ericaceae		147
Rhynia major	Rhyniaceae		6, 7
Ribes petraeum	Grossulariaceae	Rock Currant	184
Robinia pseudoacacia	Fabaceae	Black Locust,	79, 118
Rorippa stylosa	Brassicaceae	Cresse	95
Rosa arvensis	Rosaceae	Field Rose	123
Rosa sp.	Rosaceae	Rose	123, 192
Rubus fruticosus	Rosaceae	Blackberry	73
Ruscus aculeatus	Ruscaceae	Butcher's Broom	141
Salix breviserrata	Salicaceae		135
Salix glaucosericea	Salicaceae		110
Salix herbacea	Salicaceae	Dwarf Willow	135, 149
Salix purpurea	Salicaceae	Purple Osier	146
Salix retusa	Salicaceae	Stunted Willow	151, 152
Salix sp.	Salicaceae	Willow	55
Salix viminalis	Salicaceae	Cyprus Sage	126, 134
Salvia verticillata	Salicaceae	Whorled Sage	133
Sambucus racemosa	Caprifoliaceae	Red-berried Elder	71
Schinus piliferus	Anacardiaceae		35, 146
Senecio inaequidens	Asteraceae	Narrow-leaved Ragwort	148
Senecio keniodendron	Asteraceae	Giant Senecio	197
Sequoiadendron giganteum	Taxodiaceae	Mammoth Tree	191
Sigillaria sp.	Lepidendrodendrales		10
Sigillaria saulii	Lepidendrodendrales		10
Sonchus leptocephalus	Asteraceae	Fennel-leaved Sow-Thistle	56
Sonneratia alba	Lythraceae	Apple Mangrove	200
Spiraea stevenii	Rosaceae	Steven's Meadowsweet	147
Stipa sp.	Gramineae	Feather Grass	29
Taeniophyllum filiforme	Orchidaceae		202
Tapinanthus oleifolius	Loranthaceae		201
Taxodium distichum	Taxodiaceae	Bald Cypress	200
Taxus baccata	Taxaceae	Common Yew	76, 77, 86, 87, 129
Thalictrum aquilegifolium	Ranunculaceae	French Meadow-rue	133
Thamnobryum alopecurum	Bryales	Fox-tail Feather-moss	70
Thamnopteris schleichedalii	Osmundales		12
Thuja plicata	Cupressaceae	Western Red Cedar	189
Thymus serpyllum	Lamiaceae	Wild Thyme	133
Tilia cordata	Tiliaceae	Small-leaved Lime	127
Tofieldia calyculata	Liliaceae	Scottish Mountain Asphodel	41
Trochodendron aralioides	Trochodendraceae	Wheel Tree	191
Tsuga mertensiana	Pinaceae	Mountain Hemlock	183
Ulmus glabra, Ulmaceae	Wych Elm		179
Ulmus minor, Ulmaceae	Field Elm		151, 167
Vaccinium myrtillus	Ericaceae	Bilberry	35, 140
Vaccinium oxicoccus	Ericaceae	Small Cranberry	188
Vaccinium uliginosum	Ericaceae	Bog Whortleberry	126
Viburnum lantana	Caprifoliaceae	Wayfaring Tree	75
Viscum album	Loranthaceae	White-berried Mistletoe	51, 178
Vitis vinifera	Vitaceae	Grape Vine	46
Welwitschia mirabilis	Welwitschiaceae		21
Withania adpressa	Solanaceae		145
Xanthorrhoea sp.	Xanthorrhoeaceae	Grass Tree	174, 175, 196
Yucca elata	Agavaceae	Soaptree Yucca	83
Yucca sp.	Agavaceae		93
Ziziphus lotus	Rhamnaceae	Lotus Tree	99
Zosterophyllum renanum	Zosterophyllales		8

LIST OF SPECIES

Subject Index

Abrupt growth reduction ...113, 125, 163, 165, 176, 177
Abrupt growth release 113, 161
Abscission 64-67
Absent ring 125
Acrotonic 192
Adult bark 104
Adult crown 100
Adult leaves 101
Adult xylem...102
Adventitious root 127, 188
Adventitious shoot..126
Aerenchyma 11, 200
Ageing, ontogenetic..43
Ageing, physiologic..43
Air conducting canal 11
Amphitonic 192
Anastomosis 172
Ant plant 199
Aphid 111
Apical meristem 40, 43
Arid...............................21, 29, 198, 145, 148, 189, 202
Astrablue 2, 117
Avalanche 52, 54, 55, 128, 180
Axial parenchyma..36, 37, 80

Bark, see phellem...46
Bark-shedding 65
Barrier zone...61, 62, 180, 205
Basitonic 192
Below-ground stems 192
Bent rays 47, 114, 158, 212, 213
Blue-stain fungus 206
Blue-green algae...18
Bog...53, 161, 181, 188
Boreal forest 22, 28
Bottle tree..198
Branches 9, 30, 59, 96, 97, 156,
168, 169, 172, 191, 195, 202
Branching pattern 194, 199
Brown rot 206, 207
Bud activity 192, 199
Bud-scale scar 64, 118, 125
Buttressed stem..198

Callous tissue, callus 110, 111, 121, 158, 159,
176, 177, 184, 185
Cambium 10, 16, 34, 37, 40, 41, 43, 47, 49, 61, 63,
70, 71, 85, 86-89, 91, 127, 170, 176, 188
Canada balsam..217
Carboniferous......................... 4, 10-12, 16, 22, 210, 211
Carbonization...208
Cell collapse...110, 160, 161, 177
Cell contents 34, 217
Cell death 50, 121
Cell elements.. 26, 27, 34 - 37
Cell expansion..89
Cell types 26, 27, 34-37
Cell wall 8, 34, 44, 56
Cell wall growth 6, 34, 56, 57, 112
Cellulose fibers...164
Charcoal...205, 208, 209
Chernobyl 164, 165
Chestnut blight 113, 179
Chronologies 1
Climate, arid..29, 145
Climate, boreal 28, 147
Climate, tropical...28, 144, 148

Climate, temperate 28, 146
Climatic regions 28, 144
Cockchafer 163
Collateral vascular bundle 74, 84
Collapsed cells 47, 89, 90, 91, 110, 139, 160
Compartimentalization...37, 62
Competition 134, 154, 155
Compression 212
Compression wood....................... 36, 52-54, 124, 125,
128, 129, 180, 181
Concentric vascular bundle 70, 71
Cone trace 121
Coppice...156
Cork 43, 46, 48, 49, 65, 86-88, 92, 93, 104,
195, 175, 202
Cortex 9, 10, 40, 43, 48, 49, 61, 88, 89, 93, 104,
123, 140, 141
Cracks 46, 153, 159, 182
Cross dating..1
Crown shape 183
Crushed cell walls 182
Crystal druses 37
Crystal needles 85
Crystal sand 85, 87
Cushion plant 196
Cuticula 37

Decapitation...170
Decomposition.....................................58, 60, 202, 206, 207, 213
Deer scar 185
Defense mechanisms..37, 61, 62,
Defoliation 162, 164, 167, 174
Deformed wood..212, 213
Dendrochronology 1
Density variation 160
Desert...29, 145, 148, 189, 198
Devonian 4, 6, 7, 8, 16, 44
Dichotomous branching 194
Die-back 188
Diffuse porous 134, 135
Dilatation 46, 53, 57, 105, 132
Discoloration..206
Displaced rings..129
Drainage 160, 161
Drought 158, 160, 193
Ducts...110
Dutch Elm Disease 179
Dwarf shrub 62, 88, 89, 111, 128, 133, 135, 137,
147, 149-152

Earlywood ....23, 37 and many others
Eau de Javelle 217
Eccentric stem 128
Eccentricity..52
Elm disease..179
Endodermis 72
Epidermis 7, 9, 37, 41, 93, 194
Epiphyte 191, 197
Epitonic 192
Evolution 3-26
Exposed roots 33, 98, 99
Extinction 4

False ring..138
Felling date..1
Fern ....12 – 14. 64, 71, 72
Fern tree 12, 13

Subject Index

Fibres, thick-walled ...7, 15, 51, 112, 122, 134, 160, 168
Fibres, thin-walled 9, 47, 72, 93, 98, 112, 125, 136, 146, 162, 168
Fire scar..177, 178
Forest fire 174, 196
Frost crack..159
Frost rib..159
Frost ring...110, 158
Fruit scar...65
Fungi..60, 111, 189, 206

Gallery..204
Gelatinous fibers 55
Grafting...172, 190
Grazing..122, 196
Green woody stems..202
Growth, bipolar...32
Growth forms..30, 100, 134
Growth increase...113, 161
Growth reduction 113-115, 125, 163, 165, 177
Growth ring boundary ....21, 23 and many others
Growth ring in bark...47, 86
Growth, unipolar..32
Gum duct..111, 177

Hail scar...184
Haustorium 178, 201
Hazel growth..129
Heartwood..50, 60, 190
Heartwood substances...................37, 61, 76, 78, 109
Hedge...113, 169, 190
Helical thickening..37, 77
Hemicryptophytic herb........25, 53, 71, 74, 95, 127, 132
Herbivory...166, 167, 199
Hydroid..70
Hypocotyl...193, 195, 199
Hypotonic..118

Imbalance...52, 180, 181
Indistinct ring boundaries 138
Injury..62, 63, 110, 184, 185
Insect gallery..204,205
Insects..111, 162, 197, 204, 205
Internal optimization..58
Intra-annual bands..108, 109

Juvenile bark...104
Juvenile crown..100
Juvenile leaves..101
Juvenile xylem..102

Kino vein...177

Larch bud moth...162
Late frost..110
Latent shoot...127,128
Latewood ....23, 37 and many others
Latewood tracheid...150
Latewood vessel 150-152
Leaf base..12, 41, 175, 196
Leaf form..101
Leaf scar 10, 64, 66
Liana..90, 197
Libriform fiber..34, 78, 103
Light lack..136, 154
Light ring..112
Lignification 55, 56, 122, 168

Loess..189
Longevity...43
Longitudinal symmetry 189
Long shoot................................101, 103, 104, 119, 122

Macerate..35
Male flower trace 121
Mangroves..200
Microscopic preparation..213
Micro slides..217
Microtome..214, 216
Minerals..17, 210
Missing ring..150
Mistletoe 51, 178, 201
Mouse scar...185

Needle age...121
Needle trace..121
Nuclear radiation...165
Nuclei 34, 36, 47, 48, 50, 51

Oil duct..88,

Paleozoic..4
Parasites 178
Parenchyma, apotracheal..80
Parenchyma, axial..34
Parenchyma, intraannual 80
Parenchyma, paratracheal................................80, 140
Parenchyma, vasicentric...140
Peck mark...185
Perforation, ephedroid..81
Perforation, scalariform......................................44, 81
Perforation, simple..44, 80
Periderm................. 46, 48, 49, 84, 89, 92 ,93, 104, 120, 121, 184
Permafrost..113, 189
Permian...4, 22
Petrification..208
Phellem 46, 48, 49, 64-66, 92, 93
Phellogen 48, 48, 66, 92, 93, 123
Phenols...79
Phloem................. 9, 15, 34, 35, 47-50, 70-73, 86-89, 104, 109, 114, 115, 191
Phloem-tapping mistletoe.....................................201
Photography..217
Phylloclades...141, 202
Phyllodes...202
Phylogeny..26
Pitcher plant..197
Pith................. 11, 19, 33, 40, 73-75, 120, 122, 126, 140, 142 ,173, 195
Pith fleck..111
Pith stem..11
Pits, bordered 19, 21, 24, 35, 45, 211
Pits, simple..35, 45, 57, 70
Pitting, araucarioid..211
Pitting, biseriate..23
Pitting, uniseriate..23
Polarized light..111, 217
Pollarding...168
Pollution..164, 165
Polystele..72
Preparation..216, 217
Primary wall..36, 207
Protection systems..60
Protostele..72, 73

Pruning	168, 169
Radial section	42
Radial cracks	159
Ray, aggregate	161
Ray dilatation	49, 53
Ray, heterogeneous	81
Ray, homogeneous	81
Ray pits	45, 77
Ray tracheid	45
Ray, uniseriate	78
Reaction wood	54
Resin	60
Resin ducts in cortex	61
Resin ducts, traumatic	163, 171, 185
Resin ducts in xylem	61, 163, 165, 181
Resin pocket	183
Rhaphides	85
Rhizome	6, 84, 132, 134, 135, 138, 142, 188
Ring, false	138
Ring, frost	110, 158
Ring-porous	31, 150, 151, 167, 169
Rockfall	182
Root	32, 96, 98, 99, 140
Root, exposed	33, 98, 99
Rosette plant	195
Safranin	217
Salt	197, 200
Salt tree	108
Sapwood	50, 60, 78, 153, 188
Savannah	28, 177, 195, 198, 199
Scar, hidden	63, 177, 179
Sclereid	34, 57, 71, 88
Seasonal climate	31, 82, 146, 147, 147, 198
Seasonal growth	31, 56, 82
Secondary growth	39-62
Secondary growth, monocotyledons	83, 85
Secondary wall	36
Semi-ring porous	143, 135, 146, 151
Senescence	188
Shade avoidance	197
Shedding needles	10, 64-66, 121
Shedding twigs	64-67
Shock cracks	182
Shoot	32, 74, 118, 126, 140
Short shoot	118, 120, 124, 125
Sieve area	9, 35
Sieve plate	35, 85
Sieve tube	46, 47, 70-72, 85-87, 89, 91, 115
Silurian	4
Siphonostele	72, 72
Soft rot	207
Species diversity	4
Spine	123, 199
Spiral grain	189
Spiral thickening	37, 77
Spruce bud worm	162
Staining	217
Stele	72, 73
Stem thickening	46
Steppe	29
Stomata	7, 9, 194
Storage parenchyma	198
Structural diversity	192
Structural variability wood	132, 134, 136, 195
Stump	168, 170, 171

Successive cambia	19, 90
Succulent	142, 194, 202
Summer drought	160
Suppression	136, 154
Tangential section	42
Tangential resin ducts	177, 185
Tannins	37, 109
Tension wood	55, 181
Termites	204
Thorn	122
Timberline	22, 147, 158, 159, 176, 183
Tracheid	7, 9, 17, 19, 20, 23, 24, 34, 37 and others
Transversal section	42
Traumatic resin ducts	185
Tree age	188, 189
Tree ring	37 and many others
Tree size	190, 191
Twig abscission	64-67
Tylosis	36, 177
Vascular bundle	7, 9, 11, 13, 15, 31, 41, 70-75, 83-85, 91, 123, 132, 141, 194
Vegetation zone	28, 29
Vessel frequency	32
Vessel-less	53
Vessels, flame like	78, 134
Vessels, radial groups	101, 151
Vessels, solitary	146
Wavy ring boundaries	129, 137
Wedging ring	128
White rot	207
Woodpecker	185
Xylem, see vascular bundle	102
Xylem-tapping mistletoe	201
Year without summer (1816)	2